P[illegible] Hope

"A great practical book that I hope will be read by many people in all walks of life, even those who still doubt human-induced climate change. No technology even imaginable can restore nature's past healthy functioning over the Earth's greatest land areas—its vast grasslands. The solutions Courtney illustrates can and will do what is required, and he tells the story well."

—Allan Savory, president and founder of the Savory Institute

"As anyone paying attention now knows, we will be facing numerous new challenges in our agriculture and food system in the near future. The most important 'journey' we all need to make in preparing for that future is, as Courtney White points out, to restore the biological health of our soil. The hopefulness in Courtney's journey comes from his demonstration of the practical ways in which we can accomplish this task. Anyone interested in the future of food should read this remarkable, heartwarming book."

—Frederick Kirschenmann, author,
Cultivating an Ecological Conscience: Essays from a Farmer Philosopher

"*Grass, Soil, Hope* takes us on a journey from one fascinating topic—and one inspirational, hardworking individual—to another. The exciting concept of 'carbon farming,' which Courtney White clearly articulates, both in theory and with practical examples, could revolutionize our entire approach to environmental restoration. If widely applied, these techniques *would* reverse climate change, and reestablish health to the land, to ourselves, and to our communities. This is an important book that is filled with hope."

—Larry Korn, translator and editor of Masanobu Fukuoka's
The One-Straw Revolution and *Sowing Seeds in the Desert*

"Courtney White's book offers refreshing insights on 'climate-smart' agriculture. In an era when farmers and ranchers are often vilified for environmental disruptions, this analysis gives an optimistic contrast: It's a well-grounded practical outlook of the win-win outcomes of management practices by ranchers who are good stewards of soil carbon."

—L. Ann Thrupp, PhD, executive director,
Berkeley Food Institute, University of California, Berkeley

"Courtney White's journey was sparked by a question: What if we looked at carbon not just as a 'pollutant,' but from the standpoint of its role as the building block of life? What he found across the country and abroad were farmers, ranchers, and scientists who are working with the carbon cycle to build soil, restore ecosystems, and bolster productivity—in short, embracing life to generate more life. At once plain-spoken and radical, this book promises to stir up hope even among those made cynical by relentless bad news. White has made the case for hope. Whether this is turned to action is up to us."

—Judith D. Schwartz, author, *Cows Save the Planet*

"Courtney White employs a masterful blend of storytelling and science to communicate a most hopeful message: that building healthy soils—in some surprising and creative ways—can help solve our food, water, and climate challenges all at the same time. The carbon-capturing farmers, ranchers, and conservationists whose work White so elegantly describes form the vanguard of a new movement of regenerative production that deserves society's attention and support. Inspiring, thought-provoking, energizing, and—at bottom—full of hope."

—Sandra Postel, freshwater fellow, National Geographic Society

"This delightful diamond of a book is a tour-de-force that covers the story of carbon from the Big Bang to your backyard. At a time when environmental narratives have become gloomy, this book is a breath of optimism exhaled with practical recommendations for moving carbon from the air back into the soil, for the health of the planet and every creature on it."

—Fred Provenza, professor emeritus,
Department of Wildland Resources, Utah State University

"*Grass, Soil, Hope* is a wonderfully accessible account of the promise of soil and agriculture for a better climate and better future."

—Thomas E. Lovejoy, professor of environmental science and policy,
George Mason University, and senior fellow, United Nations Foundation

"*Grass, Soil, Hope* takes the reader back to earth's beginnings to help illustrate the vital role of carbon in sustaining life and then gives real-life, real-time examples of agricultural practitioners who are using creativity and common sense to grow food, restore watersheds and wildlife habitat, and, yes, sequester lots of carbon."

—William McDonald, fifth-generation cattle rancher;
founder and director of the Malpai Borderlands Group

"Grass, soil, hope: Three simple words with the power to tackle society's most challenging problems. A ray of sunshine, converted by grass into carbon and stored in the soil, represents the possibility of a brighter future. An empowering and uplifting read!"

—Gabe Brown, owner, Brown's Ranch, Bismarck, North Dakota

"This is a book to read for many reasons: to learn about the Earth's carbon cycle; to glimpse ways 'conservation' is evolving, especially in the semi-arid West; and to understand the future of ranching and sustainable agriculture. It's also a book to read if you want to be infused with hope and inspired to play a broader role in the face of climate change. For many of us who think about ways to create a more resilient world for future generations, it pays to think more about carbon. This book will get you started."

—Jonathan Overpeck, co-director, Institute of the Environment;
professor of geosciences and atmospheric sciences, University of Arizona, Tucson

Grass, Soil, Hope

ALSO BY COURTNEY WHITE

Revolution on the Range: The Rise of a New Ranch in the American West
Knowing Pecos: A Small History of a Big Place
The Indelible West: Photographs 1988–1998

Grass, Soil, Hope

A Journey through Carbon Country

Courtney White

FOREWORD BY
MICHAEL POLLAN

Chelsea Green Publishing
White River Junction, Vermont

Project Manager: Hillary Gregory
Project Editor: Benjamin Watson
Copy Editor: Eric Raetz
Proofreader: Nancy Ringer
Indexer: Lee Lawton
Designer: Melissa Jacobson

Printed in the United States of America.
First printing June, 2014.
10 9 8 7 6 5 4 3 2 1   14 15 16 17 18

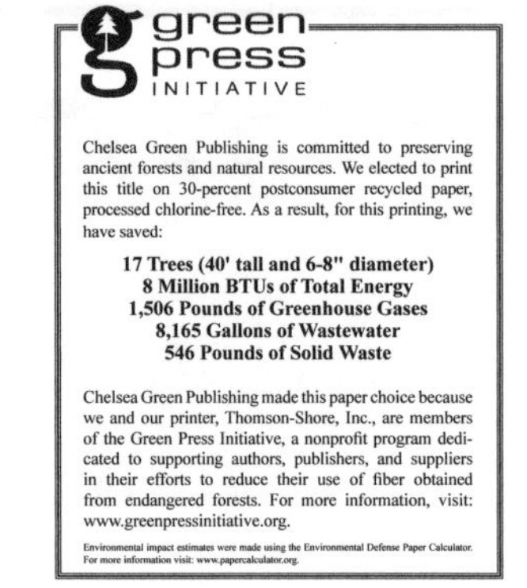

Our Commitment to Green Publishing

Chelsea Green sees publishing as a tool for cultural change and ecological stewardship. We strive to align our book manufacturing practices with our editorial mission and to reduce the impact of our business enterprise in the environment. We print our books and catalogs on chlorine-free recycled paper, using vegetable-based inks whenever possible. This book may cost slightly more because it was printed on paper that contains recycled fiber, and we hope you'll agree that it's worth it. Chelsea Green is a member of the Green Press Initiative (www.greenpressinitiative.org), a nonprofit coalition of publishers, manufacturers, and authors working to protect the world's endangered forests and conserve natural resources. *Grass, Soil, Hope* was printed on FSC®-certified paper supplied by Thomson-Shore that contains at least 30% postconsumer recycled fiber.

Library of Congress Cataloging-in-Publication Data

White, Joseph Courtney, 1960–
Grass, soil, hope : a journey through carbon country / Courtney White.
pages cm
Includes bibliographical references and index.
ISBN 978-1-60358-545-3 (pbk.) — ISBN 978-1-60358-546-0 (ebook)
1. Soil conservation—Case studies. 2. Sustainable agriculture—Case studies. I. Title.

S623.W47 2014
631.4′5—dc23

2014000290

Chelsea Green Publishing
85 North Main Street, Suite 120
White River Junction, VT 05001
(802) 295-6300
www.chelseagreen.com

This book is dedicated to my heroes
Aldo Leopold, Wallace Stegner, and Wendell Berry
and to the new agrarians who stand on their shoulders.

CONTENTS

FOREWORD

Hope in a book about the environmental challenges we face in the twenty-first century is an audacious thing to promise, so I'm pleased to report that Courtney White delivers on it. He has written a stirringly hopeful book, and yet it is not the least bit dreamy or abstract. To the contrary, *Grass, Soil, Hope* is deeply rooted in the soil of science and the practical work of farming.

Grass, Soil, Hope is at the same time a challenging book, in that it asks us to reconsider our pessimism about the human engagement with the rest of nature. The bedrock of that pessimism is our assumption that human transactions with nature are necessarily zero-sum: for us to wrest whatever we need or want from nature—food, energy, pleasure—means nature must be diminished. More for us means less for it. Examples of this trade-off are depressingly easy to find. Yet there are counterexamples that point to a way out of that dismal math, the most bracing of which sit at the heart of this book.

Consider what happens when the sun shines on a grass plant rooted in the earth. Using that light as a catalyst, the plant takes atmospheric CO_2, splits off and releases the oxygen, and synthesizes liquid carbon–sugars, basically. Some of these sugars go to feed and build the aerial portions of the plant we can see, but a large percentage of this liquid carbon—somewhere between 20 and 40 percent—travels underground, leaking out of the roots and into the soil. The roots are feeding these sugars to the soil microbes—the bacteria and fungi that inhabit the rhizosphere—in exchange for which those microbes provide various services to the plant: defense, trace minerals, access to nutrients the roots can't reach on their own. That liquid carbon has now entered the microbial ecosystem, becoming the bodies of bacteria and fungi that will in turn be eaten by other microbes in the soil food web. Now, what had been atmospheric carbon (a problem) has become soil carbon, a solution—and not just to a single problem, but to a great many problems.

Besides taking large amounts of carbon out of the air—tons of it per acre when grasslands are properly managed, according to White—that process at the same time adds to the land's fertility and its capacity to hold water. Which means more and better food for us. There it is: a non-zero-sum transaction.

This process of returning atmospheric carbon to the soil works even better when ruminants are added to the mix. Every time a calf or lamb shears a blade of grass, that plant, seeking to rebalance its "root-shoot ratio," sheds some of its roots. These are then eaten by the worms, nematodes, and microbes—digested by the soil, in effect, and so added to its bank of carbon. This is how soil is created: from the bottom up.

To seek to return as much carbon to the soil as possible makes good ecological sense, since roughly a third of the carbon now in the atmosphere originally came from there, released by the plow and agriculture's various other assaults, including deforestation. (Agriculture as currently practiced contributes about a third of greenhouse gases, more than all of transportation.) For thousands of years we grew food by depleting soil carbon and, in the last hundred or so, the carbon in fossil fuel as well. But now we know how to grow even more food while at the same time returning carbon and fertility and water to the soil. This is what I mean by non-zero-sum, which is really just a fancy term for hope.

It has long been the conventional wisdom of science that it takes eons to create an inch of soil (and but a single season to destroy it). This book brings the exceptionally good news that this conventional wisdom no longer holds: with good husbandry, it is possible to create significant amounts of new soil in the course of a single generation. The farmers and the scientists who are figuring this out are the heroes of *Grass, Soil, Hope*.

The book takes the form of a travelogue, a journey to the grassy frontiers of agriculture. Some of these frontiers White finds in the unlikeliest of places: on Colin Seis's "pasture cropping" farm in

Australia, where annual grains are seeded directly into perennial pastures; on John Wick's cattle ranch in Marin County, California, where a single application of compost has roused the soil microbiota to astonishing feats of productivity and carbon capture; in the tenth-of-an-acre "edible forest" that Eric Toensmeier and Jonathan Bates planted, according to the principles of permaculture, right behind their house in Holyoke, Massachusetts. Each of these chapters constitutes a case study in what is rightly called "regenerative agriculture." Taken together, they point the way to a radically different future of farming than the one we usually hear about—the one in which, we're told, we must intensify the depredations (and tradeoffs) of agriculture in order to feed a growing population. Courtney White's book points to very different idea of intensification—one that also brings forth more food from the same land but, by making the most of sunlight, grass, and carbon, promises to heal the land at the same time. There just may be a free lunch after all. Prepare to meet some of the visionaries who have mastered the recipe.

MICHAEL POLLAN
Berkeley, California
December 2013

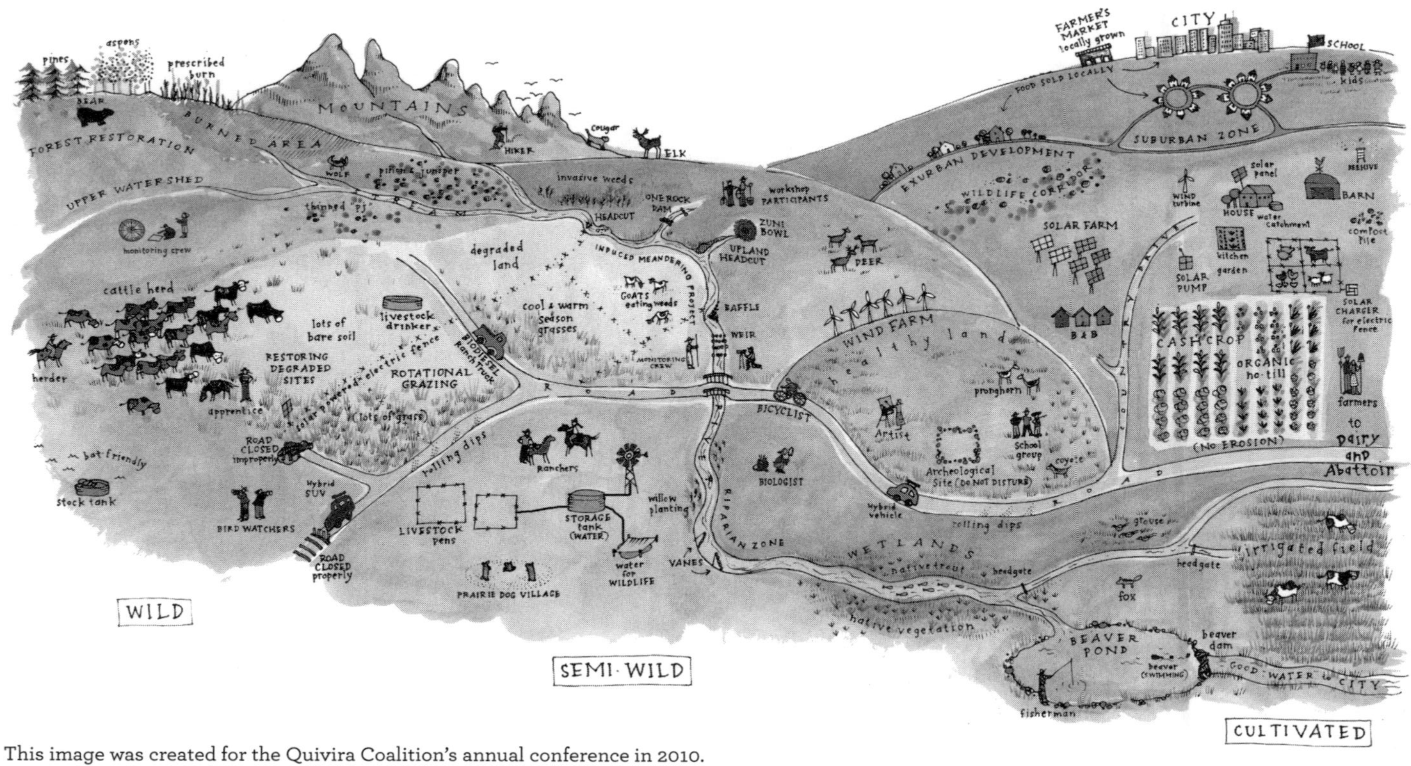

This image was created for the Quivira Coalition's annual conference in 2010.
Concept by Courtney White. Artwork by Jone Hallmark.

PROLOGUE

This is the story of how I came into Carbon Country.

I'm a former archaeologist and Sierra Club activist who became a dues-paying member of the New Mexico Cattle Growers' Association as a producer of local, grass-fed beef.

For a boy raised in the suburbs of Phoenix, Arizona, during the heyday of sprawl, fast food, and disco music, this was a bewildering sequence of events. I grew up surrounded by cars, malls, concrete, transplanted cacti, and copious amounts of air-conditioning. The closest I came to livestock were the horses my parents owned for trail-riding purposes. Cattle? Local food? Sustainability? I had no clue. Even when I became active with the Sierra Club in the mid-1990s after a move to Santa Fe, New Mexico, my conservation work was highly conventional. I lobbied for new wilderness areas, protested clear-cut logging in national forests, and helped publish a citizen's guide to fighting the environmental damage caused by hard-rock mining. I led activist outings, organized letter-writing campaigns, testified in public hearings, and fought a cynical assault on environmental regulation at the time called "takings" legislation. When I had time to think about livestock grazing at all, it wasn't in a positive light.

This all changed in 1997 when I cofounded the nonprofit Quivira Coalition with a rancher and a fellow conservationist. I did it because the constant brawling between environmental activists and loggers, ranchers, and other rural residents had dispirited me. No one was winning; everyone and everything was losing, especially the land. Even worse was the negative energy employed by all parties involved—attacking each other in the media, pointing fingers in meetings, filing lawsuits in court, even threatening physical violence. There had to be another way. When I met a rancher who not only did things differently on his land but sought a different relationship with environmentalists, I decided it was time to give peacemaking a chance.

With Quivira, we waded into the middle of the grazing wars in a deliberate attempt to create a "third position" outside the continuum of combat. We called it the New Ranch—a meeting place "beyond wrongdoing and rightdoing," to quote the poet Rumi, where people interested in innovative ideas and fruitful dialogue would have a place to meet, talk, listen, and learn.

It wasn't just talk, however. The New Ranch meant managing land differently, including moving livestock around in ways that mimicked the natural behavior of migratory herds of wild grazers. New Ranchers operated on the principle that the natural processes that sustain wildlife habitat, biological diversity, and functioning watersheds are the same processes that make land productive for livestock. It wasn't just a theory—it worked in practice, as I saw over and over on ranch after ranch. The key was *land health*: the degree to which the integrity of the soil and ecological processes of rangeland ecosystems are sustained over time. I learned that before land can sustainably support an added value—such as livestock grazing, hunting, recreation, or wildlife protection—it must be functioning properly at a basic ecological level. This included healthy water, mineral, and energy cycles, flowing round and round from the soil to plants and animals and back again.

With Quivira, my conservation work became highly collaborative, with a focus on improving land health, promoting progressive cattle management, implementing creek restoration projects, and repairing damaged relationships. My Sierra Club experience had taught me a hard lesson: that the missing piece of the conservation puzzle was the positive role that people could play. Environmental problems, I came to understand, were as much about social and economic relationships as they were about the environment, thus requiring economic solutions to go along with ecological ones. I learned this by listening to the many heated confrontations between activists and ranchers and loggers over the years. Conservation, I saw, meant prudence, care, good stewardship, and trust as much as it meant passing laws, enforcing regulations, and establishing new parks. That's why I chose a quote from farmer and author Wendell Berry as Quivira's motto: "We cannot save the land apart from the people; to save either, you must save both." Saving both became the mission of the Quivira Coalition.

Over time, our collaborative work grew to include an annual conference, a ranch apprenticeship program, a capacity-building collaboration with the Ojo Encino chapter of the Navajo Nation, numerous publications, a ton of workshops, and lots of creek restoration projects—including a long-running project in northern New Mexico on behalf of the Rio Grande cutthroat trout. By our calculation, at least 1 million acres of rangeland, 40 linear miles of creeks, and countless people have directly benefited from Quivira's collaborative efforts across the Southwest.

The membership in the New Mexico Cattle Growers' Association happened in 2006 when 49 heifers were delivered to Quivira's 36,000-acre Valle Grande ranch, located on a national forest near Santa Fe. They were the first installment of what would become a 124-head herd of heifers, plus three Corriente bulls, all under our "Valle Grande" brand and our management. Shortly thereafter, an invitation to join the Cattle Growers' Association arrived in our office. We filled out the form, wrote a check, and mailed it back. And just like that, this former Sierra Club activist became a dues-paying cattle rancher!

Our plan was to sell grass-fed beef in Santa Fe, joining the rapidly growing local food movement, and use the revenue to pay for conservation activities on the ranch. For a while it worked. Thanks in part to best-selling books by Michael Pollan and Barbara Kingsolver, grass-fed beef was an easy sell to customers. In 2008 I had the honor of traveling to Turin, Italy, as a delegate to Slow Food's biennial Terra Madre gathering as a producer of local, grass-fed beef. It was an experience that changed my outlook on conservation. Food made people smile, I saw, binding us together. It was positive energy at work again, reminding me that the only lasting change is the one that comes from the heart.

Unfortunately, our happy little world began to unravel in the fall of 2008. The financial meltdown on Wall Street, a product of huge amounts of negative energy (greed), triggered a cash-flow crisis for Quivira and other nonprofits as the stock portfolios of foundations and donors shrank dramatically. Grass-fed beef suddenly looked expensive to customers as well. All of this put our Valle Grande ranch in financial jeopardy, calling to mind the old joke: "How do you make a small fortune in ranching? Start with a big one." We didn't start with *any* fortune, big or small. Soon we were forced to sell our cattle herd to pay the bills. Eventually we had to sell the ranch too. This was a big disappointment personally, but I vowed to put our experience to good use somehow.

Meanwhile, I had begun to fret about the Big Picture.

It started in the spring of 2006, during a fund-raising trip to New York City. Rummaging in an airport bookstore for something to read on the outward leg of my journey, I came across James Kunstler's best-selling cautionary tale *The Long Emergency: Surviving the Converging Catastrophes of the Twenty-First Century.* Curious, I plucked the book from the rack and flipped it over to survey the promotional blurbs, reading how the author "graphically depicts the horrific punishments that lie ahead for Americans for more than a century of sinful consumption and sprawling communities, fueled by the profligate use of cheap oil and gas." Yikes! Then I thought, "Oh come on, how bad could things be?" I handed the clerk fifteen dollars to find out.

Bad enough to refocus Quivira's mission, as it happened.

At our annual conference in 2007, Wendell Berry said that "we are not walking a prepared path," in response to a question from the audience about the difficulties posed by the twenty-first century. In other words, to meet new challenges we need to blaze a new trail. That suggested unexplored country ahead, which is after all what the word *quivira* originally designated on old Spanish maps of the New World. After some thought, I decided this new trail was building ecological and economic *resilience,* which the dictionary defines as "the ability to recover from or adjust easily to misfortune or change." In ecology, it refers to the capacity of plant and animal populations to handle disruption and degradation caused by fire, flood, drought, insect infestation, or other disturbance. The word also has a social dimension. Ranching, for instance, is the epitome of resilience, having endured centuries of cyclical drought, low cattle prices, and other challenges.

Resilience is also an important concept for those of us who live in cities, as I had learned the previous winter when a major snowstorm shut down both highways into Albuquerque, New Mexico, isolating the city. In a story for the local newspaper, a reporter asked how long it would take for the shelves of Albuquerque's grocery stores to be emptied of food. His answer: six days. That's not very resilient. What about other challenges, I wondered, such as our supply of fresh water? Was it resilient for the long run? Were we?

Realizing that the times were changing, in the fall of 2007 we added the words "build resilience" to Quivira's mission statement. In doing so, I realized that I was now a long way from the grazing wars of the 1990s—not to mention the suburbs of Phoenix.

There was a lot to learn in this new country. Take climate change. It wasn't on our radar screen at all in 1997, but a decade later it had become a major concern. As I learned, the rising content of heat-trapping gases in the atmosphere, carbon dioxide (CO_2) especially, poses a dramatic threat to life on Earth. Here's a graph from the Scripps Institution of Oceanography at the University of California–San Diego, which pretty much sums it all up.[1] It's a scientific projection of CO_2 (in parts per million).

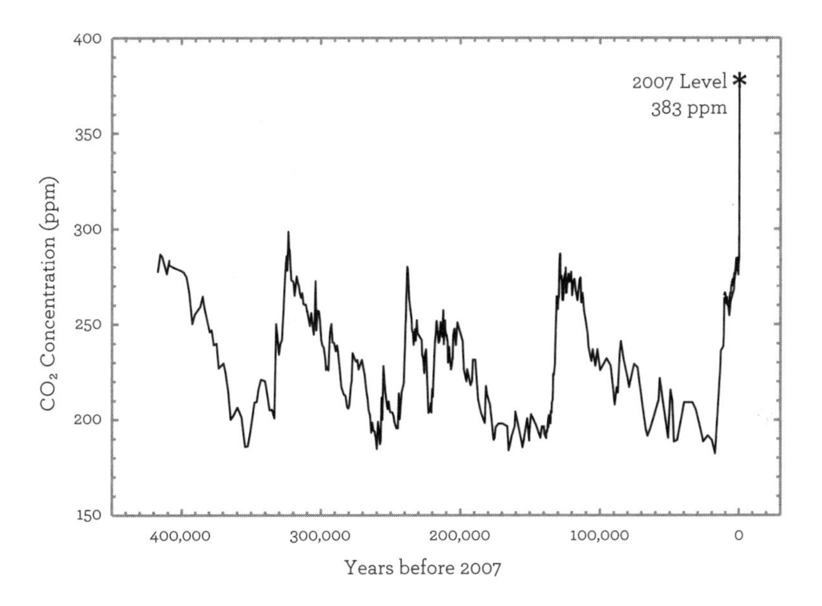

CO_2 over the past 420,000 years

In 2013, the CO_2 level rose above 400 ppm for the first time in five million years, according to researchers, and it is on a trajectory to reach 600 to 700 ppm by the end of the twenty-first century, with all sorts of bad consequences, unless we act quickly. Double yikes!

Something needed to be done, but what? In 2009, I found a partial answer in an op-ed written by James Hansen, the director of NASA's Goddard Institute for Space Studies and the nation's top climate scientist. Reducing the carbon dioxide content of the atmosphere back to 350 ppm, he said, is imperative to preserve a habitable planet. "If we cut off the largest source of carbon dioxide—coal—we have a chance to bring CO_2 back to 350ppm," he wrote, "and still lower through agricultural and forestry practices that increase carbon storage in trees and soil."[2] Cool! I thought to myself. But what did he mean by "carbon storage"?

An explanation arrived a month later when a publication came across my desk from the Worldwatch Institute titled *Mitigating Climate Change through Food and Land Use.* Its authors, Sarah Scherr and Sajal Sthapit, wrote that for political, technological, and economic reasons, the only possibility for large-scale removal of greenhouse gases from the atmosphere currently is through improved ecosystem function, climate-friendly livestock practices, conserving land, and restoring degraded watersheds.[3] I did a mental double take. That sounded like the work of the Quivira Coalition!

The miracle cure is called *photosynthesis.* As Scherr and Sthapit pointed out, plants naturally pull CO_2 out of the air and convert it into soil carbon, where it is safely stored for long periods of time in the ground unless disturbed—by plowing, for instance. This process has been going on for billions of years, and all it requires is sunlight, green plants, water, nutrients, and soil microbes. It creates a simple equation: more plants and deeper roots = less CO_2 in the atmosphere.

It's more complicated than that, of course. But here's the really exciting part: if land that is bare, degraded, tilled, or monocropped can be restored to a healthy condition, with properly functioning carbon, water, mineral, and nutrient cycles, and covered year-round with a diversity of green plants with deep roots, then the added amount of atmospheric CO_2 that can be stored in the soil is potentially high.

Globally, Scherr and Sthapit wrote, soils contain about three times the amount of carbon that's stored in vegetation and twice the amount stored in the atmosphere. Since two-thirds of the earth's land mass is grassland, additional CO_2 storage in the soil via better management practices, even on a small scale, could have a huge impact. Grasslands are also home to two billion people who depend on livestock—an important source of food and wealth (and culture) to much of the earth's human population. Both these animals and their human stewards could be mobilized for carbon action.

This made huge sense to me, so I called Scherr and invited her to speak at Quivira's annual conference in the fall of 2010, which I had titled "The Carbon Ranch." The purpose of the event was to describe the many ways by which food and stewardship can be used to build soil, store carbon, and fight climate change. I told her I was determined to explore this exciting country and spread the good news. When she agreed to make a presentation, I began calling up other carbon pioneers, eventually assembling an exciting lineup of speakers. But then a thought struck me: Where was I going? Climate? Carbon? Where had we wandered off to?

I decided we needed a map.

I sat down one morning at my dining room table and began sketching on a sheet of paper. I drew every joyous, sustainable, resilient, regenerative, land-healing, relationship-building, climate-mitigating, local food–producing activity I could pull from my experience, putting them into a single mythical landscape. I sketched (badly) cattle-herding ranchers, weed-eating goats, bat-friendly water tanks, creek-restoring volunteers, land health–monitoring crews, fish-friendly wetlands, grass-fed beef businesses, no-till farms, and on-site renewable energy projects. Then I added cities, schools, farms, beavers, wolves, bird-watchers, kitchen gardens, wildlife corridors, compost piles, and more. I intentionally left out boundaries, including property lines, political divisions, and geographical separations. There was no distinction on my map between public and private land, or between wild country and nonwild. It was all one map—all one vision in which wolves, cattle, bats, organic farmers, biologists, artists, foxes, fish, cities, and ranchers all worked together and got along.

When I was done, I sat back and studied my map. I knew this place. It was the land I had been exploring for years—except it wasn't. I hadn't considered it from a carbon perspective before. It felt like a new country, ripe for further exploration. But where would I go? What would I discover? Were there actual on-the-ground solutions to the rising challenges of the twenty-first century? If so, was there an answer to an increasingly anguished question being asked by Americans of all stripes: what can I do to help?

I knew a few things going in:

Carbon is key. It's the soil beneath our feet, the plants that grow, the land we walk, the wildlife we watch, the livestock we raise, the food we eat, the energy we use, and the air we breathe. Carbon is the essential element of life. Without it we die; with too much we suffer; with just the right amounts we thrive. A highly efficient carbon cycle captures, stores, releases, and recaptures biochemical energy, making everything go and grow from the soil up. In the last century or so, however, the carbon cycle has broken down at critical points, most importantly in our soils, which have had their fertility eroded, depleted, and baked out of them by poor stewardship. Worse, carbon has become a source of woe to the planet and its inhabitants as excess amounts of it accumulate in the atmosphere and oceans. It's all carbon. Climate change is carbon, hunger is carbon, money is carbon, politics is carbon, land is carbon, we are carbon.

We don't have to invent anything. Over the past thirty years, all manner of new ideas and methods that put carbon back into the soil and reduce carbon footprints have been field-tested and proven to be practical and profitable. We already know how to graze livestock sustainably, grow organic food, create a local food system, fix creeks, produce local renewable energy, improve water cycles, grow grass on bare soil, coexist with wildlife, and generally build resilience into the land and in our lives.

It's mostly low-tech. It's sunlight, green plants, animals, rocks, mud, shovels, hiking shoes, windmills, trees, compost, and creeks. Some of the work requires specialized knowledge—such as herding livestock or designing an erosion-control structure in a creek—and some of it has high-tech components—such as solar panels or wind turbines—but most of Carbon Country can be easily navigated by anyone.

Lastly, **you're on the map too**. Everyone is, whether you live in a city, go to school, graze cattle, enjoy wildlife, grow vegetables, hike, fish, count grasses, draw, make music, restore creeks, or eat food—you're on the map. You live in Carbon Country. We all do. It's not a mythical land; it exists.

This is what I knew—and all that I knew. Surveying the map, I realized that there were specific questions that needed answers: (1) Was it actually possible to significantly increase the amount of CO_2 in soils via land management practices and thus impact climate change, as the experts suggested? (2) What were the range of activities that sequestered carbon in soils? (3) Was it practical to scale up sequestration practices and their cobenefits in ways that would address rising challenges in the twenty-first century? (4) What paradigms would need to be shifted to make this work possible? (5) What were the best incentives to make all of this work economically? (6) Who was going to do all this new work?

It wasn't clear, so with my rough map in hand, I set out to explore this new land. Here's what I discovered.

1

ESSENCE

This is a story about carbon and hope.

It begins in the gentle, rolling hills surrounding the tiny town of Nicasio, in Marin County, California, where I thought I had made a wrong turn into Ireland. It was March, 2010, and the hills were a vibrant, heart-pounding grassy green, the sort of green that resonates someplace deep within us like an ancient memory. If you're from a dry place—New Mexico in my case—green like this means you stop for a photograph. I parked near the church, climbed out of the car, and marveled at the emerald grass all around me. What is it about green that lifts our spirits? We're not ruminants after all, to whom an ungrazed field of grass represents breakfast, lunch, and supper. Nor am I a pastoralist, with herding stick in hand, to whom a fresh field is the equivalent of a daily blessing or a bank account. Instead, I was just a guy with a camera and a rental car traveling to a ranch to learn something new under the sun, waylaid by green. Lots of green. What is it about this color that feels like hope? Something eternal, I thought, as I lifted the camera to my eye. Something important. I snapped a photo.

Without knowing it, the click of the camera signaled the start of this story.

I was in Nicasio to visit John Wick, an environmentalist turned rancher turned carbon pioneer. John and his wife, Peggy Rathmann, are owners of the Nicasio Native Grass Ranch, a 540-acre property they purchased in 1998 in order to get out of San Francisco, where John worked in construction management and Peggy built a successful career as an author of children's books. They bought the ranch, which borders the Point Reyes National Seashore to the west, to be close to nature and indulge a passion for native plants. John is an energetic, slender, bespectacled man who radiates charm and passion. It is easy to imagine John throwing himself into his work, whether managing a construction project in the city or managing a small ecosystem in the country. At the time, John considered himself to be an environmentalist, which is why when they moved to the ranch the first thing he did was cancel a grazing lease with a neighbor and kick the cows off their land. They assumed that livestock were incompatible with nature, native plants in particular, as they had been led to believe. They subscribed to the conventional zero-sum paradigm of the movement: that conservation could only advance as far as the cattle retreated.

They don't believe that anymore.

As John tells the story, once the cows departed, their ranch quickly turned into a weedy mess, with all sorts of exotic and noxious plants popping up all over, threatening the very ecosystem John and Peggy were determined to protect. In response, they turned to the standard remedies: pulling, spraying, mowing, and burning the weeds with a vengeance. But none of it made a difference. The objectionable plants kept coming back, stronger than ever. In something like desperation, John turned to an acquaintance, Jeffrey Creque, an agroecologist who was part of the management team at a nearby organic olive farm.

Visiting the ranch, Jeff gently told John and Peggy that the weeds were a symptom of their land's poor health, not the cause. He urged them to think holistically. Pulling, mowing, burning, and especially spraying, he told them, were activities that *suppressed* life. Instead, as landowners they should be *encouraging* life on the ranch by working with nature and its processes, not against it. For example, during an ecological inspection of the ranch, Jeff pointed out that a great deal of native grasses lived among the noxious weeds. Instead of asking "What can I do to kill something?" they should have been asking, "What can I do to help something flourish?" In this case, how could they encourage the native grasses to grow and proliferate, outcompeting the weeds? These were two totally different questions with two different courses of action.

It was all about attitude.

This is bigger than weeds and grass, of course. We humans tend to focus on the *problem* rather than the *potential* of a particular situation, too often dwelling on the negative rather than exploring the positive. Where John saw trouble on his land, Jeff saw possibility. Where John saw mowing, burning, and spraying as necessary to removing something he didn't like, Jeff saw these activities as an assault on life in general. Jeff pointed out, for example, that spraying herbicide kills a lot more than weeds—it kills the microbial life in the soil too, which is bad because microbes are essential to the health and vigor of *all plants*. Jeff's advice to John and Peggy was to flip their goals from negative to positive, work on the potential of their land, and manage their ranch for health and life—instead of death.

Now, if this sounds all "very California"—the granola state—it's not. In my experience across the American West and beyond, in a wide variety of geographies, cultures, and economies, when people employ positive energy for positive purposes—to produce healthy food, for instance, or build community—instead of negative energy for negative ends—such as killing things—a world of possibility opens up. I've seen it happen over and over. John and Peggy understood all this intuitively and agreed to do things differently on the ranch, including thinking about the Big Picture. What, they asked Jeff, was the first step?

Bring back the cattle, he said.

Wait, John objected, weren't cows also life suppressors? Didn't they kill things too, by overgrazing the grass and trampling the soil? Weren't they the reason the weeds were there in the first place? No, Jeff replied, the problem wasn't the cattle themselves, but rather the way they were managed. Herbivores and grasses have coexisted with each other for at least sixty-six million years in North America, Jeff pointed out. Grass and grazers depend on each other. Obviously cattle are recent arrivals, he told John, but they're still ruminants, and they can do a pretty good job as proxies for wild grazers, such as bison, if they are allowed to behave like them. The key is management, and the goals of the managers. If the goals are positive—managing land for health and life, for instance—then cattle can be a powerful tool for good. However, if the goals are negative—treating nature as if it were an exploitable commodity—then livestock can be destructive. Think of a hammer: it can be used to either build a house or hit someone on the head. Same tool, different goals. The tool of grazing, Jeff said, could be used to revitalize the nutrient cycle in a ranch's soils, especially if that land has been depleted by poor management. It happens via the "disturbance" effect of herbivory—eating, walking, and defecating—all of which can stimulate grass growth when properly managed. If mismanaged, however, overgrazing by cattle can easily damage the land's health.

Same animal, different results.

John acknowledged that he understood Jeff's point about encouraging the grass to outcompete the weeds. He liked the whole idea, in fact—everything except the cattle part. It was a paradigm thing, he admitted. "Couldn't go there," he told me with a smile as we drove across the ranch after my arrival. "There had to be another way. I thought—why not elk?" He decided to approach the managers of the neighboring national seashore, hoping to "borrow" the resident elk herd to graze his ranch. It seemed like a win-win proposition: he needed grazers, and elk always need fresh grass. That was the theory. The reality was different. John quickly ran into a thicket of bureaucratic obstacles to his plan, not the least of which was an objection from the California Department of Fish and Wildlife, which said, in so many words: we've never heard of such a thing! Besides, they reminded him, there were rules.

"No," in other words.

"It was okay," John told me. "I was disappointed at first, but then I realized it was for the best. Elk, after all, are not very good at following instructions." John needed the grazing to be targeted and managed, which meant he needed the animals to follow a grazing plan about where to go and how long to stay. "Wild animals aren't great with that," he admitted, "but cows are."

John let his paradigm go.

After consulting with Jeff and other experts, John developed a plan for the ranch that divided the property into eighty-five paddocks (on 540 acres) based on grass quality and soil type. This allowed him to exercise a large degree of control over when and where the animals went and what and how much they ate—all easily accomplished with portable electric fencing. The next step was to find the bovine workforce. He swallowed his pride and called his rancher neighbor. "Could the cows come back?" John asked politely. He promised to do all the work and provide all the cattle necessities, including the grass. All his neighbor had to do was pay a fee. His neighbor agreed and they signed a contract. Soon the cattle were on the job, moving through the paddocks, eating grass. Sometimes they stayed in a paddock for as little as four hours before moving to the next one, but never longer than three days. John followed the plan diligently, working hard to survive his crash course in cattle ranching, but he still harbored some initial skepticism about the ultimate outcome, he told me. His doubts, however, evaporated rapidly.

"I was amazed at the quick results," he said, "especially as the native grasses and wildflowers rebounded." Life spread across the land. "It works," he said matter-of-factly as we bounced down a road. "And I knew it worked the moment I saw eagles return to the ranch," he continued, smiling again. "That was cool."

And that's how John Wick became a rancher-environmentalist.

John's story doesn't stop here, however. The next step on his journey is even more remarkable—and the reason why I had rented a car, driven to Nicasio, and stopped to take a picture of the Irish-colored hills. John's next stop was carbon—and this is where the truly hopeful part of the story begins.

It started with a talk. On November 4, 2007, John drove up the road to hear permaculture specialist Darren Doherty speak about soil carbon—the stuff that makes life thrive underground. What he heard changed John's life. Like many people, John and Peggy worry about the buildup of greenhouse gases in the atmosphere. This buildup began with the invention of agriculture nine thousand years ago, expanded with the advent of fossil fuel combustion during the Industrial Revolution, and has dramatically accelerated in recent decades. The main culprit is carbon dioxide, a colorless and odorless gas that has existed on Earth in small but varying quantities for billions of years. In many ways, CO_2 has been given a bad rap. After all, it is essential for the maintenance of life on the planet. As a heat-trapping element in the atmosphere, CO_2 acts as a "blanket" for all life on the planet, and without it Earth would plunge into a deep freeze. Just as critically, all green plants require CO_2 to live. Plants "inhale" the gas through their pores, break the C apart from the O_2 via the miracle of photosynthesis, keep the carbon, and "exhale" the oxygen back into the air (thankfully for us mammals). In other words, CO_2 is a *good* gas. But like any substance, it's only good in proper amounts. Scientists warn us that there is too much CO_2 in the atmosphere currently, creating a crisis situation for life on Earth due to rising temperatures. In fact, there haven't been this many blankets wrapping the planet at any time in the past five million years. For life to continue in some semblance of normality, levels of CO_2 must come down, scientists say—way down.

This isn't a new problem. Over the eons, greenhouse gases in Earth's atmosphere, which include methane and nitrous oxide, have risen periodically to levels unhealthful for life before falling again. In the past, Earth handled the problem of excess carbon dioxide by employing its *sinks*—vast natural "sponges" that absorb CO_2 and store it safely for varying degrees of time. There are four major sinks on the planet: the atmosphere, oceans, forests, and soils. The first three are becoming problematic today due to the unprecedented rate at which CO_2 is being pumped into the air. According to scientists, the atmosphere is now officially "full" of CO_2 and overflowing worryingly; oceans are rapidly "filling up" with the gas and may "top off" in only a few decades; and forests and other vegetation, while potentially able to soak up a lot of CO_2 (if we manage them properly), can't provide long-term storage due to their habit of burning up and dying, releasing the gas back into the air. All of which brings us to soil. The organic material in soil, the stuff that makes soil dark, rich, and good-smelling—called *humus*—is largely carbon. Where did the bulk of this carbon originate? As CO_2.

This is where Doherty's talk comes in.

According to John, Doherty reported that recent research indicated that a mere 2 percent increase in the organic content of the planet's soils, particularly in its grasslands, *could soak up all the excess carbon dioxide in the atmosphere within a decade.* The hairs on John's neck stood up. Soils around the planet, Doherty continued, have been mismanaged for centuries, resulting in the depletion of their original organic content and thus their capacity to soak up CO_2. But now, in a huge irony, all these depleted soils are available to start absorbing all that troublesome atmospheric CO_2 again—if we rebuild their organic content through better land management. Our mission, Doherty said, is to build topsoil and save the planet.

John was stunned. Could it be true? Could it be that simple?

John knew that Jeff Creque and his coworkers had *doubled* the carbon content of the soil on the 500-acre olive farm, from 2 to 4 percent, in less than ten years. With a Ph.D. in rangeland ecology and decades of experience as an organic farmer, Jeff had been hired by the managers of the olive farm in 1997 to address the question of what to do with the waste products from the new olive oil mill. To answer this, Jeff embarked on a soil-building strategy that included:

1. applying lots of compost, made on-farm from olive mill waste + livestock manures + landscaping debris harvested on the ranch;
2. avoiding tillage via the maintenance of a permanent cover crop beneath the olive trees;
3. performing seasonal rotational grazing of sheep through the orchard; and
4. restoring riparian areas, to address downcutting gullies on the property.

"Olive oil is like butter," Jeff told me. "It's produced from the current season's photosynthetically derived carbon. If the farm exports only oil, it essentially removes nothing permanently from the soil. By avoiding tillage and returning all residuals to the land, the olive oil agroecosystem takes in more carbon from the atmosphere than it emits. Done well, olive oil production can be an essentially permanent, regenerative form of agriculture."

Data backs him up. Dozens of soil samples are taken every year from all over the olive farm and sent to a laboratory for analysis. While results have shown year-to-year fluctuations in the organic matter content of the soil, mostly due to weather and sampling variables, the trend has been clear: upward. In fact, after ten years the carbon content of the soil began hovering around 4 percent. This means the farm is likely sequestering more CO_2 than it did back in 1997; certainly the soil is more productive and it's holding more water.

Jeff believes that what they accomplished can be repeated elsewhere, as he explained to John. After all, millions of tons of organic waste—including food, grass clippings, branches, and manures—go into landfills every year across the nation, where they rot and produce a substantial amount of methane, a potent greenhouse gas. Why not compost them instead, Jeff suggested, and divert them to farms, ranches, and other types of open space where they could help build soil and reduce greenhouse gas emissions simultaneously? There's a carbon cost to hauling this material around by truck, of course, but that could be offset by the land's increased ecological productivity and all the CO_2 sequestration it will provide.

It all sounded practical and exciting to John, but soak up *all* the excess CO_2 in the atmosphere? That was crazy talk. Had to be!

Crazy or not, the wheels in John's head were turning. As soon as he returned home he began to read about the carbon cycle, which is the process by which carbon dioxide flows out of the atmosphere and into the soil via photosynthesis and green plants as organic carbon, then back out again into the air via decomposition and respiration. Round and round in a perpetual cycle, sustaining nearly all life on Earth. It wasn't a complete cycle, however. Some of the carbon stayed in the soil, having exited the plants' roots in order to feed the microbial life there, where it can remain lodged for decades if left undisturbed. In other words, Doherty was right, at least in principle: excess CO_2 not only could be pulled out of the air by a natural process, but it could be stored safely in the soil for long periods of time. This is called *sequestration*, which the dictionary defines as the process of "safekeeping, withdrawing, or seizing for the purposes of placing in custody." What if we became carbon sheriffs, John thought, seizing excess CO_2 and placing it into the custody of soils for a lengthy "jail term," where it would feed plants and promote life? Talk about win-win!

John's wheels really began to turn. Three questions leapt to mind: Could farming and ranching actually play a significant, positive role in reducing excess CO_2? Could the practices that sequester CO_2 in the soil also improve land health—and profits? Could this be the basis of a new carbon economy?

John decided to find out.

However, before he could start, he had to answer a technical question: How do you actually *measure* carbon in soils? And, more importantly, how do you measure *changes* in the carbon content of soils as a result of management? There were lots of different methods, as it turned out. Which was best? It wasn't simply an academic issue because John wanted to make the data useful to ranchers, farmers, and their customers—as well as to convince potential skeptics, of which he assumed there were many. Personally, John suspected that he was building up the carbon content on his land as a result of his new job as a rancher-environmentalist, but he didn't have any numbers to support his hunch.

A visit to the Internet gods wasn't helpful, so he turned to Whendee Silver, a biogeochemist and professor of ecosystem ecology at the University of California–Berkeley. Biogeochemistry studies the relationship between chemical elements, including carbon and nitrogen, and living things such as biological systems in the soil. Generally, it focuses on cycles, or the ways elements, nutrients, energy, and water flow through the natural world. The carbon cycle, for example, is a major area of research today, and Silver is an expert. Surely, John thought, she would know the answers to his questions.

She didn't—not at first, anyway. She was intrigued, however. Her laboratory at Berkeley was already halfway to what John needed, since one of its goals was to study the impacts of deforestation, reforestation, grazing, and other land-use practices on carbon sequestration and biogeochemical cycling. What John wanted to know, however, put research questions squarely into the realm of real-world management. Silver saw this as an opportunity to meld science and agriculture so that each could inform the other in useful and practical ways. Toss in the "save the world" part and it was an opportunity that was too good to pass up. She not only decided to try to answer John's questions, she also agreed to be the lead scientist in what suddenly blossomed into a full-fledged research endeavor.

And so, in early 2008, the Marin Carbon Project was born.[1]

The principal goal of the project is to explore the value of local soil carbon sequestration in rangelands, including private ranches and publicly owned open spaces, in order to provide ecological and agricultural benefits to rural communities. Accomplishing this goal, the researchers knew, would be exciting, daunting, and complicated. They started by building a collaborative team and then set out to gather baseline scientific data on soil carbon across the state. Then they turned to the question of setting up test plots on John and Peggy's ranch and collecting the data.

Soon, they were the subject of a National Public Radio story, reported by Christopher Joyce, titled "Scientists Help Ranchers Wrangle Carbon Emissions." It's how I first heard about the project, while doing the dishes in our kitchen. The story ran on December 10, 2009, just as the critically important United Nations conference on climate change was beginning in Copenhagen, Denmark. The news from Denmark was already rotten, I thought, including figurative shoving matches between delegations and a great deal of foot-dragging by the big industrialized nations. A comprehensive agreement to reduce greenhouse gas emissions looked like an increasingly impossible goal. The NPR story caught my attention because Joyce used the words "carbon ranching" to describe the efforts of John and his colleagues at the Marin Carbon Project. I had begun researching the idea of carbon sequestration, which I had labeled a *carbon ranch*, in soils the previous summer, so I paused in my dish duties to listen to the following story.

Some people in Marin County, California, may already have a partial solution. They call it "carbon ranching." The idea was hatched by scientists who are trying to coax carbon dioxide out of the air and into cattle pastures. Proponents of the idea say if it proves effective, the practice could be used around the world.

Whendee Silver is a soil scientist at the University of California, Berkeley. If soil is the earth's skin, then Silver might be considered its dermatologist.

"What we're interested in doing out here is figuring out how much carbon is added to the soil and how much carbon is lost," she says.

Soil and the plants that grow in it depend on carbon. Essentially, carbon dioxide is plant food, and Silver wants them to eat more. To encourage the uptake of carbon dioxide, Silver has spread compost over these plots of pastureland. The compost is a mix of plant clippings and animal manure, the same kind you might put on your garden at home.

The compost, she says, "increases plant growth, it actually also lowers the temperature a little bit, so the soil doesn't get quite as hot and it doesn't stimulate as much microbial activity."

Her experiment seems to be going well. The grass here is visibly taller, which means there is more carbon in the plant, which also means more food for cows. Ranchers like that part. But those microbes she mentions complicate the process. Soil is full of them, and when they eat plants, animals, and bugs, they emit carbon dioxide into the air. So Silver's composting has to work a balance between supercharging carbon-consuming plants—without beefing up carbon-producing microbes.

Scientists are experimenting with grassy pastures like this one in California to increase how much carbon dioxide the pastureland captures from the air. So far, the grass in the composted plots grows so well that it captures 50 percent more carbon from the air than grass in the untouched plots. And the soil is taking up almost all the carbon in the compost—carbon that likely would have gone up into the atmosphere if it hadn't been added to the pastureland. Silver is now measuring exactly how much that is.

"Grasslands, because they are in these dry regions, have really, really high root biomass, and it tends to go pretty deep. These plants are looking for water and that's what builds that dark, organic rich soil and that carbon-rich soil," says Silver.

Silver thinks composting could work for thirty years before the soil is saturated with carbon. During that time, Silver says ranchers could see a payoff of sorts for their work. "Hopefully, they'll be able to participate in a carbon market, where we can quantify how much carbon is being stored on the land, and we can sell that as a carbon offset," she says.

That idea intrigues John Wick, a rancher who owns grazing land where Silver is conducting her experiments for the Marin Carbon Project. "Now I think about carbon in everything I'm doing, and it's completely changed my life. This whole ecosystem down there is alive, I mean, up until this point it was just dirt to me, something I pushed around with my bulldozer," says Wick.

This all sounds complicated, and it is. But as negotiators at the Copenhagen climate meeting struggle with ways to reduce carbon dioxide in the atmosphere, storing carbon in soil and plants may start to look like an attractive option.[2]

Wow. The idea of a carbon ranch suddenly didn't seem so crazy after all!

Unfortunately, the climate negotiators in Copenhagen failed to produce a binding agreement of any sort. All follow-up meetings have likewise failed. This means greenhouse gas emissions will continue to rise for quite some time into the future. Suspecting this possibility, I flew to California in the spring of 2010 to meet John. I wanted to know more about the Marin Carbon Project and its potential for mitigating climate change. I was also intrigued by John's statement, "Now I think about carbon in everything I'm doing, and it's completely changed my life." What did he mean by that?

Most people hardly give carbon a passing thought, except for the bad press it has received in the climate change context. The main message people get is that carbon, in the form of hydrocarbons such as coal and oil, is one bad dude—a ne'er-do-well at best, a villain at worst. That's why there is a lot of talk among experts and activists about a "postcarbon economy" as well as a "war on carbon." Carbon has become another enemy, apparently, that has to be defeated. As I quickly learned, however, John completely disagreed with carbon's bad rap. It was the attitude thing again: he saw carbon as an opportunity, not as a problem. And not just any opportunity—one of the most important in our lifetimes. It wasn't simply a matter of "locking up" carbon in the soil, he told me, but employing an element essential to life on Earth to do all sorts of good things, such as growing food. Carbon was a tool, like the proverbial hammer, that could be used for good or evil depending on one's goals. The villainous side of carbon had been well documented and publicized. The purpose of the Marin Carbon Project, John said, was to demonstrate the element's heroic side.

All of which raised a question in my mind: what is carbon, actually? I had no clue. Upon my return home from California, I decided to dig. Here's what I learned.

Carbon is the graphite in our pencils, the diamond in our rings, the oil in our cars, the sugar in our coffee, the DNA in our cells, the air in our lungs, the food on our plates, the cattle in our fields, the forest in our parks, the cement in our sidewalks, the steel in our skyscrapers, the charcoal in our grills, the fizz in our sodas, the foam in our fire extinguishers, the ink in our pens, the plastic in our toys, the wood in our chairs, the leather in our jackets, the battery in our cars, the rubber in our tires, the coal in our power plants, the *nano* in our nanotechnology, and the life in our soils.

Carbon is everywhere. It is the fourth most abundant element in the universe, the fifteenth most abundant element on Earth, and the second most abundant element in the human body, after oxygen. Carbon is present in all known life-forms. It can be found dissolved in all water bodies on the planet. Most of it is stored in rocks. Carbon is abundant in the sun, stars, comets, meteorites, and the atmospheres of most planets (the atmosphere of Mars, for instance, is 96 percent carbon dioxide). Carbon is a basic element—number six on the periodic table of elements, between boron and nitrogen. It exists in many inorganic compounds (gases, rocks, liquids) and in all organic ones. Half of the dry weight of most living organisms is carbon.

Carbon is stardust. It first formed in the interior of stars, not long after the Big Bang, and then scattered into space as a result of supernova explosions. Over time, it coalesced into second- and third-order star systems such as ours, as well as planets, comets, and other heavenly bodies. Eventually, it coalesced into us. Joni Mitchell had it right in her song "Woodstock": we are stardust.

Carbon is promiscuous. It forms more chemical compounds than any other element, with almost ten million compounds discovered to date—a tiny fraction of all that are theoretically possible. Carbon especially likes to bond with other small atoms, including other carbon atoms. This makes it capable of forming long chains of complex and stable compounds, which is why it is found in so many different forms on Earth.

Carbon is history. In antiquity, it was called *carbo*, which is Latin for coal. Carbon was known to the earliest humans as soot and to the earliest civilizations as diamonds. The Romans made charcoal for cooking and the Amazonians made *terra preta* (biochar) for burying to improve their nutrient-poor soils. No one knew these were all the same element until 1772, when French chemist Antoine Lavoisier pooled money to buy a diamond, which he placed in a closed glass jar. He focused the sun's rays on the diamond with a special magnifying glass and saw the diamond burn and disappear. There was no water left behind. The jar was filled with carbon dioxide—gram for gram. This was similar to what happened with charcoal. He called this common element "carbone." In 1779, scientist Carl Scheele did the same thing with graphite—and carbon's reputation took off.

Carbon is energy. The bonds in long chains of carbon hold a lot of energy, and when they are broken that stored energy is released. This makes carbon an excellent source of fuel, both in machines and in living creatures. Coal, oil, natural gas (methane), tar sands, bitumen, and everything in between are all hydrocarbons, a highly stable and yet easily combustible bond between hydrogen and carbon. Refrigerants, lubricants, solvents, plastics, chemical feedstocks, and other types of petrochemicals are all hydrocarbons. The world would grind to a halt without carbon.

Carbon is life. It exists in every organic life-form. Life is impossible without it. When combined with water, it forms sugars, fats, alcohols, and terpenes. When combined with nitrogen and sulfur, it forms amino acids, antibiotics, and alkaloids. With the addition of phosphorus, it forms DNA and RNA, the essential codes of life, as well as ATP, the critical energy-transfer molecule found in all living cells. The carbon atom is the essential building block of life. Every part of your body is made up of chains of carbon atoms, which is why we are known as "carbon-based life-forms." Chemically, we're just a bunch of inert compounds. What breathes life into us? The answer is the relationship between the molecules of energy and nutrients, fueled by carbon and water.

Carbon is hope. Because where there's life, there's hope.

Carbon is also a cycle. The carbon in the atmosphere, the oceans, the trees, the soils, us, and everything else is constantly in motion, flowing in a giant circle from air to land and back to air again in an unending, closed loop. The law of the conservation of matter says that in a closed system matter can be neither created nor destroyed. It can only cycle and recycle. Earth has been a closed system almost from its origin, with only solar energy, an occasional electromagnetic pulse from the sun, and stray bits of asteroids and comets entering the atmosphere from space. What's here today has always been here, including carbon, whose total amount is essentially the same as it was when Earth formed. The ancient Greek philosophers understood all this intuitively, proclaiming "nothing comes from nothing." Epicurus wrote that "the totality of things was always as it is now, and always will be." Nothing can be created or destroyed. This observation was explained scientifically by none other than Lavoisier, who discovered that, although matter may change its form or shape—a diamond into gas—its mass always remains the same.

So it is with carbon. And what carbon does is cycle—a process essential to life on Earth. It's a carefully regulated process too, so that the planet can maintain critical balances. Call it the Goldilocks Principle: not too much carbon, not too little, but just the right amount. For instance, without CO_2 and other greenhouse gases, Earth would be a frozen ball of rock. With too many greenhouse gases, however, Earth would be an inferno like Venus. "Just right" means balancing between the two extremes, which helps to keep the planet's temperature relatively stable. It's like the thermostat in your house. If it gets too warm, the cycle works to cool things off, and vice versa. Of course, the planet's thermostat gets overwhelmed at times, resulting in periods of rapid warming or cooling (think ice ages). No matter what happens, the miraculous carbon cycle keeps working, scrubbing excess CO_2 out of the atmosphere, or adding more if necessary. The carbon cycle never sleeps.

Who does all this regulatory work? Two quick answers: green growing plants and evolution. Photosynthesis is the rapid process by which carbon is transferred from sky to soil. It's what makes the Goldilocks Principle tick. Evolution is the methodical process by which life changes over succeeding generations—what lives, what dies, which population expands, which one contracts. It keeps the Goldilocks Principle ticking over time—long periods of time. The two work in concert. The quantity of carbon in the environment influences the course of evolution, and vice versa. The effects of an excessive buildup of CO_2 in the atmosphere, for example, will impact the fate of generations of living things. Carbon and evolution interact and adjust to one other, regulating and responding in a sophisticated dance. Carbon chooses the music, if you will, while evolution dictates the steps in a planetwide choreography. It is a dance with a profound effect on audience members.

During the Carboniferous Period of Earth's history, for instance, which began 350 million years ago and lasted for fifty million years, the music was turned up very loud. A potent combination of swampy terrain, warm temperatures, high humidity, and unprecedented levels of oxygen caused an explosion of life across the planet. Insects grew to huge sizes. Modern-looking fish evolved. Birds, reptiles, and mammals began to lay eggs on solid ground for the first time—in a fateful evolutionary leap. It was the vegetation, however, that really went wild. As the period's name implies, massive amounts of carbon-bearing trees grew during this time, many of which toppled into swamps when they died, becoming entombed in muck. Layer after layer of trees and muck piled up, creating over 300 million years later the rich coal seams that we recklessly exploit today for our energy.

Carbon is not the only dance on the planet, of course. Our world is full of cycles—water, energy, nutrients, nitrogen, phosphorus, and many more—each interacting with the others in complicated ways. Some cycles are short, like a song, while others are long, like a symphony or a mass. Carbon has it both ways. Its short, or fast, cycle revolves around green plants and photosynthesis—the process by which carbon is separated from oxygen, stored in roots and soils, and released back into the atmosphere via death and decomposition. Its long, or slow, cycle is geologic—what happens when carbon is released after being trapped or frozen in layers of rock for millions of years. In the case of the slow cycle, the symphony is *really* long—carbon can take between one hundred and two hundred million years to rotate fully through rocks, soil, ocean, and atmosphere.

In the slow cycle, carbon in the atmosphere combines with water vapor to form carbonic acid (in a weak solution) that falls to the ground in rain events and begins to dissolve rocks—a process known as *chemical weathering*. This process releases minerals, including potassium, sodium, calcium, and magnesium, all of which are carried by streams and rivers to the ocean. In the ocean, calcium and carbon combine to create calcium carbonate, a mineral necessary for shell-making creatures, such as corals and plankton, to grow—a key to life underwater. When these organisms die, they fall to the seafloor, where over time they become carbonate rocks, such as limestone. Then after more time (a lot more), carbon is returned to the atmosphere via volcanic activity. Ejecta flies upward into the air in the form of ash, lava, or other material. Volcanism also releases trapped carbon dioxide—and the cycle starts all over. Round and round, very slowly. If too many volcanoes go off at once and the carbon cycle becomes unbalanced due to excessive CO_2 in the atmosphere, the process of chemical weathering will help rebalance things again, but only after thousands or millions of years.

Luckily, it is the fast carbon cycle—photosynthesis—where our hope can be found.

The process by which atmospheric CO_2 gets converted into soil carbon has been going on for at least a billion years, and all it requires is sunlight, green plants, water, nutrients, and soil microbes. It's an equation: healthy soil + healthy carbon cycle = storage of atmospheric CO_2. One of the first researchers to recognize the significance of this equation was Christine Jones, an independent soil scientist in Australia, whom I had the pleasure of meeting in New Mexico not long after my visit to California. She firmly believes that the fast carbon cycle can be managed for plant, animal, and human health.

In summary, according to Jones, there are four basic steps to the CO_2/soil-carbon cycle:

Photosynthesis. This is the process by which energy in sunlight is transformed into biochemical energy, in the form of a simple sugar called glucose, via green plants—which use CO_2 from the air and water from the soil, releasing oxygen as a by-product.

Resynthesis. Through a complex sequence of chemical reactions, glucose is resynthesized into a wide variety of carbon compounds, including carbohydrates (such as cellulose and starch), proteins, organic acids, waxes, and oils.

Exudation. Around 30 to 40 percent of the carbon created by photosynthesis can be exuded directly into soil via plant roots to nurture the microbes that help plants grow and build healthy soil. This process is essential to the creation of topsoil from the lifeless mineral soil produced by the weathering of rocks over time. The amount of increase in organic carbon is governed by the volume of plant roots per unit of soil and their rate of growth. In other words, more active green leaves mean more roots, which mean more carbon exuded.

Humification. This process involves the creation of humus, a chemically stable type of organic matter composed of large, complex molecules made up of carbon, nitrogen, minerals, and soil particles. Visually, humus is the dark, rich layer of topsoil that people generally associate with stable wetlands, healthy rangelands, and productive farmland. Think of the rich soil in a productive backyard garden—that's humus. Land management practices that promote the high ecological integrity of the soil are key to the creation and maintenance of humus. Once

> carbon is sequestered as humus it has a high resistance to decomposition, and therefore can remain stable for hundreds or even thousands of years.

"Over millennia a highly effective carbon cycle has evolved to capture, store, transfer, release and recapture biochemical energy in the form of carbon compounds," Jones has written in a paper. "The health of the soil—and therefore the vitality of plants, animals and people—depends on the effective functioning of this cycle."[3]

For years, Jones has argued that soil carbon doesn't get the respect it deserves from scientists, farmers, conservationists, or the public at large. Partly that's because it is hidden underground, but mostly it's because we're just beginning to understand the critical role carbon plays in our lives and livelihoods. Economies, Jones notes, can create wealth by buying and selling valuable commodities such as gold, diamonds, and oil in a market setting, but economists and others consistently overlook the world's most frequently traded commodity: soil carbon. Almost all life on Earth depends on liquid carbon for maintenance, growth, and reproduction, she says, and would come to a stop without it. Trees, crops, grasslands, animals, humans, and just about all other living, growing things on the planet depend on the soluble carbon they get directly from other plants and animals. This makes carbon the currency for nearly all transactions between living beings.

Nowhere is this more evident than in the soil, says Jones, where a massive and intense "carbon market" takes place among plant roots, colonies of bacteria, and networks of mycorrhizal fungi, which represent a symbiotic relationship between a fungus and the roots of a plant. These fungi are the middlemen in this marketplace, brokering liquid carbon from plant roots for mineral nutrients from bacteria, protozoa, nematodes, and other soil biota. In scientific terms, this involves "bidirectional flow" in which nutrients are transported to roots by fungi at the same time that dissolved carbon moves in the opposite direction. Jones says mycorrhizal plants can transfer as much as fifteen times more carbon into the soil as adjacent nonmycorrhizal plants. Indeed, spreading the word about the good deeds of the mycorrhizal universe has become something of a passion for her.

The flip side of this good news is the extensive damage humans have done to this microbial marketplace over the millennia, principally through poor land management, including plowing, overgrazing, and the application of industrial chemicals. Many modern fertilizers and biocides, Jones notes, inhibit microbial diversity in the soil, preventing natural carbon flow. This in turn reduces the plants' mineral uptake and thus affects mineral density in the food we eat. These practices also inhibit the process of soil humification, requiring even greater expenditure on industrial chemicals in an attempt to control the pests, weeds, diseases, and fertility problems that ensue. The overall result has been a steady depletion of carbon stocks among the world's soil marketplaces.

Jones likes to tell the story of the nineteenth-century Polish explorer and geologist Sir Paul Edmund Strzelecki, who traveled throughout southeastern Australia between 1839 and 1843, visiting farms and analyzing their soils. He wanted to know what made soils productive. To find out, he collected forty-one soil samples from farms that he thought exhibited either high or low productivity. His subsequent analysis of the samples revealed that the most important factor was the amount of soil carbon present, measured as "organic matter" in Strzelecki's day (organic matter is generally calculated to be 58 percent carbon). Of the forty-one samples, the top ten most productive had organic matter levels ranging from 11 to 37 percent (averaging 20 percent). In contrast, the organic content of the bottom ten soils ranged from 2 to 5 percent (averaging 3.7 percent). Strzelecki also discovered that the soils with the highest levels of organic matter had the highest water-holding capacity, with an eighteen-fold increase between the lowest and the highest samples.

To Jones, this was huge news. That's because the average organic content of Australia's soils today is 2 to 5 percent. Clearly, something went wrong over the 150 years since Strzelecki's visit. What went wrong, she says, was a type of agriculture imported from humid England that depleted arid Australia's soils of their carbon stocks. As she began to assemble numbers from her research, the message they revealed became clearer and clearer: the most meaningful measurable indicator for the health of the land, and thus the long-term wealth of a nation, is whether soil carbon is being accumulated or lost. If carbon is being lost, so is the economic and ecological foundation on which agriculture is based. She notes that the number of farmers in Australia has fallen 30 percent in the last twenty years and is continuing to decline. There is also reluctance on the part of young people to return to the land, indicative of the poor image and low income-earning potential of current farming practices. It suggests to her that time is running out.

"The longer we delay undertaking changes to land management," she writes, "the more soil (and soil carbon and soil water) will be lost, exposing an increasingly fragile agricultural sector to escalating production risks, rising input costs and vulnerability to climatic extremes. It's time to move away from depletion-style, high emission, chemically based industrial agriculture and get serious about grass-roots biologically based alternatives."

It is her conviction that the formation of topsoil can be breathtakingly rapid once the biological dots have been connected. The sun's energy, captured in photosynthesis and channeled belowground as liquid carbon, fuels the microbes that ignite biological activity and make minerals soluble, enabling the creation of humus in deep layers of soil. These minerals end up in plant leaves, where they elevate the rate of photosynthesis, increasing levels of liquid carbon channeled into the soil, which frees up more minerals . . . round and round. When we eat these plants or when we eat animals that have eaten them, the minerals end up in our bodies, mingling with our own stardust.

Enhancing the natural flow of carbon to soils will result in increased microbial diversity, improved nutrient cycles, expanded water-holding capacity, greater resilience, improved ecosystem health, and a more satisfying, profitable future for farmers and ranchers. It's not an impossible task, Jones insists: "When pastures are managed to utilize nature's free gifts—sunlight, air and soil microbes to rapidly form new, fertile, carbon-rich topsoil—the process is of immense benefit to farmers, rural communities and the nation. The key to subsurface management is above-ground management."

Her point is supported by research conducted by Richard Conant and two colleagues in a landmark paper published in 2001. They reviewed 115 studies from around the world, discovering that soil carbon content increased with improved land management in 74 percent of the cases. This indicated that historical soil losses can potentially be reversed, they wrote, and atmospheric carbon can be sequestered with good agricultural practices such as reduced tillage, improved fertilizer management, elimination of bare fallowing, the use of perennials in rotations, and the use of cover crops. Furthermore, within established pastures soil carbon can be increased by a variety of management techniques that have evolved to increase forage production for livestock, including fertilization, irrigation, introduction of earthworms, intensive grazing management, and sowing of favorable forage grasses and legumes. "Under certain conditions, grazing can lead to increased annual net primary production over ungrazed areas," they wrote, particularly "in areas with a long evolutionary history of grazing and low primary production."[4]

Cattle, in other words, can help heal the land while reducing the oversupply of CO_2 in the atmosphere. This, of course, was Jeffrey Creque's point—and John Wick's revelation.

Which brings me back to the Marin Carbon Project.

In the fall of 2012, Whendee Silver and her graduate student Rebecca Ryals published the results of their research on carbon sequestration in the soils of John Wick's ranch and on a separate site in the Central Valley of California. They wanted to test what would happen if a single application of composted waste from a local dairy were applied to a section of rangeland. They hypothesized that plant production would increase due to the presence of additional organic matter (i.e., carbon), but only for one year, and that this increase would be partially or wholly offset by elevated rates of greenhouse gas emissions—from soil organisms. Biologically, when bacteria in the soil are active, they consume organic matter and respire (exhale) greenhouse gases. Most of this activity takes place in a layer of soil nearest the surface, which is where the *labile* carton is located. This is the carbon that is most biologically active, which means it breaks down more quickly than carbon stored at deeper levels in the soil. From a sequestration perspective, it is the deeper, more stable carbon that is the most promising.

From a research perspective, Silver and Ryals wanted to know: Would greenhouse gas emissions rise in comparison to a companion plot that didn't receive an application of compost? Would carbon sequestration be stimulated? And would the *net effect* be one of sequestration or emission?

The test was conducted over three years at the two sites. Annual grasses, like the ones that dominate California's rangelands, grow and then die every year, which means that there is a lot of bacterial respiration going on belowground and a substantial amount of CO_2 and other gases leaking into the atmosphere. But can these grasses be "encouraged" to sequester more than they produce? To find out, the researchers applied a single layer of compost to test plots (paired with control units) at the two sites and then monitored the results over the next three years. As they expected, the manure boosted plant productivity and raised soil respiration by 20 percent (for CO_2 only, however). To their surprise, the sustained increases of plant growth lasted for three years, with no sign of diminishing effects. Even more surprisingly, data showed that the increase in plant production *significantly offset elevated soil respiration from microbial activity in five out of the six paired plots*. In other words, more atmospheric CO_2 went into the soil than came back out—a lot more. "We conclude that a single application of composted organic matter can significantly increase grasslands carbon storage and that effects of a single application are likely to carry over in time," they wrote.[5]

According to Silver, this is significant for two reasons. First, there are nearly 60 million acres of rangeland in California, and if 50 percent were available for carbon sequestration, up to 40 metric tons of CO_2 per year could be removed from the atmosphere, which is roughly equivalent to the total amount of CO_2 produced by the state's commercial and residential sectors. Second, this system creates a way to divert organic waste from landfills and dairies, thereby reducing greenhouse gas emissions from these traditional sources.

Combined, these two results could have a tremendous effect on California's contribution to climate change.

"Our results have important implications for rangeland management in the context of climate change mitigation," they concluded. "Urban and agricultural green waste is often an important source of greenhouse gas emissions. Here we show that an alternative fate for that material can significantly increase [plant productivity] and slow rates of ecosystem carbon loss."

Music to John Wick's ears.

John now had the solid numbers he needed to validate his original hunch that significant amounts of carbon from CO_2 can be sequestered in California's rangeland soils under certain conditions. This was big news of course, but for John there were two other reasons to celebrate. First, solid numbers like these will win over skeptics and light a fire under decision makers, regulators, landowners, and others to get going with carbon sequestration projects. Second, they make a strong case for the role of sustainable agricultural practices in California's burgeoning carbon marketplace, the product of climate change legislation signed into law by Governor Arnold Schwarzenegger a few years ago to implement a cap-and-trade program. This market requires hard numbers to work properly, and now John and his friends had them—opening up the possibility of a carbon economy, one of John's cherished dreams.

The numbers were music to my ears as well, because they supported claims made by soil scientists and activists that the prospect of carbon sequestration is viable and practical on a global scale, creating the possibility that it could have a substantial impact on the buildup of atmospheric carbon dioxide and thus climate change. For me, however, just as important as the data is what the experiment said about the philosophy at work. There *is* a different way to look at the world based on life instead of death, creating opportunity where before there was only despair. It *is* all about attitude—about looking at familiar things in a new light, about using old tools for new purposes, about asking the right questions and putting the answers to work.

It's about the things that nurture life—love, kindness, care, affection, experience, knowledge, laughter, liberty, family, food, and the pursuit of happiness. It's about the miraculous and the mundane, the inexplicable and the experimentally proven. It's about values and motives, dreams and fears. What sort of world do we want our children to inherit—one that is alive and vibrant, or one that is not? Do we use our tools and our knowledge and our art for positive, regenerative goals, or do we continue to kill things that we don't understand or don't care to understand? The choice is ours.

However, I've learned the hard way that solid numbers are often not enough to change attitudes or behaviors. I'm often reminded of rancher and former Quivira board member Roger Bowe, who changed the management of his family's ranch when he took over from his dad. Roger grew tons of new grass, brought springs back to life, and increased the ranch's financial bottom line significantly as a result—and had the numbers to prove it. As a trained economist, he knew precisely what the numbers meant and how to present them to others, especially fellow ranchers. Unfortunately, it didn't make much of a difference. His neighbors would come over to the ranch, observe the abundant grass, examine his healthy cows, listen to Roger talk about the innovative way he moves cattle across his land, hear the numbers, go home, and change nothing. They liked the old way—despite the numbers—even if it meant going bankrupt in the end. Economists call this an example of "irrational" economic behavior. It's a story I heard repeatedly across the region, and it never failed to flummox me until I began to realize there was more at work than just numbers. There were beliefs too—often *deep* beliefs, as Roger discovered.

The answer to this conundrum is a careful, compelling blend of science and common sense. If you're trying to convince others, especially doubtful others, it's important to start with solid numbers. Whether you're trying to make a case before a scientist, an activist, a reporter, a judge, or a loan officer, your argument will be strengthened by good, hard data. It's also useful when it comes to running a ranch. "If you can't measure it, you can't manage it," is a favorite saying of Gregg Simonds, a colleague who has managed large ranches in the West innovatively and profitably for years. Without numbers, in other words, it's just guesswork. However, numbers sometimes aren't enough, especially where nature—and human nature—is involved.

What motivates people to stop killing things, for example, and learn to heal and grow life instead? Hard numbers? Peer-reviewed science? Cold economics? Certainly for some, but for many others a stronger motivation lies in the realm of ethics, morals, and human feelings. Take the motivation we feel when we see the color green or the compassion we feel toward animals. These are things that deeply resonate in us spiritually. In my experience, people don't need numbers to understand a good idea if the idea resonates with them. It's called *sense*, as in "that makes sense," and I've seen it in operation over and over—just like the way it made sense to John Wick to give cattle a try. So did the idea of sequestering carbon in soils—a hunch eventually backed up by research.

My hunch about carbon ranching had been backed up as well by all that I had seen and heard in California. I knew now that my journey into this strange and exciting land, which I began to call Carbon Country, was just beginning.

As I drove away from John and Peggy's little ranch, I took another look at the Irish-colored hills behind the Nicasio church. Everything was different. I still saw grass, but I now considered the universe of carbon it represented and the critical role it plays in the complex story of life on this amazing planet. As I drove past the hills, I recalled a quote by the poet T. S. Eliot, who wrote that "the end of all our exploring will be to arrive where we started and know the place for the first time." Thinking about carbon and climate meant seeing a familiar land again for the first time. It wasn't just the Phoenix of my youth or the unexplored territory of Quivira, but something more fundamental. Something new and old at the same time. Like us. We are stardust. Carbon is part of our essence. Its story needs to be told—and heard. It is the story of our past, our present, and our future. It is our story. It is the story of grass. The story of green.

The color of hope.

Healing the Carbon Cycle with Cattle

JX RANCH, EASTERN NEW MEXICO

On a sky-blue October day, I drove into the dry country of eastern New Mexico to visit an award-winning ranch and contemplate the carbon cycle. Although I had visited many well-managed ranches over the years, I had never looked at one through a carbon lens before, especially in the context of carbon sequestration and climate change. I was certain that the ranchers hadn't, either. The *crisis-that-shall-not-be-named* is a politically charged topic in the rural West, even among the best ranch managers, which makes the challenge of talking about carbon without talking about *you-know-what* a delicate juggling act.

On the other hand, for cattle ranchers like Tom and Mimi Sidwell, it's not necessary to bring up the topic at all. That's because healing the carbon cycle is what they do for a living. Whether it improves *you-know-what* isn't on their minds.

In 2004, the Sidwells bought the 7,000-acre JX Ranch south of Tucumcari, New Mexico, and set about doing what they know best: earning a profit by restoring the land to health and stewarding it sustainably.

As with many ranches in the arid Southwest, the JX had been hard used over the decades. Poor land and water management had caused the grass cover to diminish in quantity and quality, exposing soil to the erosive effects of wind, rain, and sunlight, which also diminished the organic content of the soil significantly, especially its carbon. Eroded gullies had formed across the ranch, small at first, but growing larger with each thundershower, cutting down through the soft soil, biting into the land deeper, eating away at its vitality.

Water tables fell correspondingly, starving plants and animals alike of precious nutrients, forage, and energy.

Profits fell too for the ranch's previous owners. Many had followed a typical business plan: stretch the land's ecological capacity to the breaking point, add more cattle when the economic times turned tough, and pray for rain when dry times arrived, as they always did. The result was the same—a downward spiral as the ranch crossed ecological and economic thresholds. In the case of the JX, the water, nutrient, mineral, and energy cycles unraveled across the ranch, causing the land to disassemble and eventually fall apart.

Enter the Sidwells. With thirty years of experience in managing land, they saw the deteriorated condition of the JX not as a liability but as an opportunity. Tom began by dividing the entire ranch into sixteen pastures, up from the original five, using solar-powered electric fencing. After installing a water system to feed all sixteen pastures, he picked cattle that could do well in dry country, grouped them into one herd, and set about carefully rotating them through the pastures—never grazing a single pasture for more than seven to ten days in order to give the land plenty of recovery time. Next he began clearing out the juniper and mesquite trees on the ranch with a bulldozer, which allowed native grasses to come back.

As grass returned—a result of the animals' hooves breaking up the capped topsoil and allowing seed-to-soil contact—Tom lengthened the period of rest between pulses of cattle grazing in each pasture from 60 days to 105 days across the whole ranch. More rest meant more grass, which meant Tom could graze more cattle—to stimulate more grass production. In fact, Tom increased the overall livestock capacity of the JX by 25 percent in only six years, significantly impacting the ranch's bottom line. The typical stocking rate in this part of New Mexico is one cow to 50 acres. The Sidwells have brought it down to one to 36 acres, and hope to get it down to one to 30 acres some day. Ultimately, Tom hopes to have the ranch divided into twenty-three pastures. The reason for his optimism is simple: the native grasses are coming back, even in dry years. Over the past ten years, the JX has seen an increase in diversity of grass species, including cool-season grasses (which grow primarily in the spring and fall), and a decrease in the amount of bare soil across the ranch. Simultaneously, there has been an increase in the pounds of meat per acre produced on the ranch.

The key to the Sidwells' success? Goals and planning. That doesn't sound terribly novel, but it's *how* the Sidwells do it that separates them from the large majority of other ranchers. They take a holistic approach, planning the management of the ranch's resources as a whole, not just parts of a whole as is traditionally done. It's the classic triple bottom line—land health, human well-being, and financial prosperity. They accomplish this by monitoring and replanning based upon what the monitoring tells them. When Tom does the annual planning for the ranch, he takes into account not only the needs of the livestock, but also the physiological needs of the vegetation and wildlife. He considers soil health to be the key to the ranch's environmental health, so he *plans* to leave standing vegetation and litter on the soil surface to decrease the impact of raindrops on bare soil, slow runoff to allow water infiltration into the soil, provide cover for wildlife, and feed the microorganisms in the soil. He also *plans* for drought, adjusting his livestock numbers *before* the drought takes off, instead of during or after the drought has set in, as is traditional.

"I plan for the drought," Tom said with a wry smile, "and so far, everything is going according to plan."

I like to think of what the Sidwells do as an *aikido* approach—go with the dryness instead of resisting it. Every fall, once the growing season is over, Tom checks his monitoring plots and evaluates how much grass he has left. Then he calculates the stocking rate for his cattle, assuming that it won't rain again until July. If it does rain or snow before then, he'll adjust the rate upward; if it doesn't, at least then he knows he can stay in business, and within the land's carrying capacity, until the monsoon rains begin. It's not the amount of rain that matters; it's how it's used when it does come. If 10 inches of rain falls on barren, eroded soils it will be less effective than 5 inches falling on grass-covered range. The first runs off, the second sinks in.

Aikido ranching.

There is an important collateral benefit to all their planning: the Sidwells' cattle are healing the carbon cycle. By growing grass on previously bare soil, by encouraging plants to send their roots deeper, and by increasing plant size and vitality the Sidwells are sequestering more CO_2 in the ranch's soil than the previous owners had. It's an ancient equation: more plants mean more green leaves, which mean more roots, which mean more carbon exuded, which means more CO_2 can be sequestered in the soil, where it will stay. Tom wasn't monitoring for soil carbon, but everything he was doing had a positive carbon effect, as evidenced by the increased health and productivity of their ranch.

There's another benefit to carbon-rich soil: it improves water infiltration and storage due to its sponge-like quality. Recent research indicates that one part carbon-rich soil can retain as much as four parts water. This has important positive consequences for the recharge of aquifers and base flows to rivers and streams, which are the lifeblood of towns and cities.

It's also important to people who make their living off the land, as Tom and Mimi Sidwell can tell you. In 2010, they were pleased to discover that a spring near their house had come back to life. For years, it had flowed at a miserly rate of 0.25 gallon per minute, but after the Sidwells cleared out the juniper trees above the spring and began managing the cattle for increased grass cover, the well began to pump 1.5 gallons a minute, twenty-four hours a day!

In fact, the water cycle has improved all over the ranch, a consequence of water infiltrating down into the grass-covered soil, rather than sheeting off erosively as it had before. This is good news for microbes, insects, grasses, shrubs, trees, birds, herbivores, carnivores, cattle, and people. Through careful monitoring, Tom is able to chart the health of his land and the effects of his decision making, which in turn influence the next round of decisions. But like an aikido master, he relies as much on his experience, intuition, and well-honed skills as on any particular set of data, which is why many ranchers like Tom call what they do the "art of rangeland management." It is part science (monitoring), part experience (planning), and part "feel" (intuition) rolled together into a skill set that is visibly successful—just like a great aikido performance.

In 2011, the Sidwells' skills were put to the test when less than 3 inches of rain fell on the JX over a period of twelve months (the area average is 16 inches per year). In response, Tom sold nearly the entire cattle herd in order to give his grass a rest. He had enough forage from 2010 to run higher cattle numbers, but asked himself, "What would a bison herd do?" They would have avoided an area in drought, he decided. It was a gamble, but it paid off in 2012 when it began raining again, although the total amount was 10 inches below normal. "It was enough to make a little grass," he told me. "We had some mortality on our grass and a lot more bare ground than before the drought, but I think the roots are strong and healthy and recovery will be quick." The Sidwells decided to purchase cattle and are now nearly 30 percent stocked and back in business. In contrast, all of their neighbors kept cattle on their land through 2011 and had to destock due to a lack of grass—just as the Sidwells began restocking!

"Grazing and drought planning are a godsend," said Tom, "and we go forward with a smile and confidence because we know we can survive this drought."

What also helped was the economics of grass-fed beef. In 2009, the Sidwells converted their beef business from a conventional feedlot-based system to an entirely grass-fed, direct-marketed operation. Grass-fed means the animals have spent their entire lives on grass—which is what nature intended for them—and *no time* in stinky feedlots, eating corn and other assorted industrial by-products. Grass-fed beef consumes less fossil fuel in its production and distribution, especially if the customers are only a short drive away from the farm, ranch, or processing facility. It also has another benefit: profitability. As an added-value food, grass-fed meat sells for as much as 50 percent more than conventional meat—if customers are willing to pay the higher premiums, which in the Sidwells' case they are. And this extra profit, even on a smaller herd, has allowed the Sidwells to make it through the dry times financially.

What the Sidwells have done on the JX is reassemble the carbon landscape. They have reconnected soil, water, plants, sunlight, food, and profit in a way that is both healing and sustainable. They did it by reviving the carbon cycle as a life-giving element on their ranch and by returning to nature's principles of herbivory, ecological disturbance, soil formation, microbial action, and good food. In the process, they improved the resilience of the land and their business for whatever shock or surprise the future may have in store.

This has important implications for the *crisis-that-shall-not-be-named*, I thought as I drove away. As the Sidwells have demonstrated, mitigation of atmospheric carbon dioxide can happen as a cobenefit of healing land, fixing the water cycle, producing food, and building resilience. Of course, Tom Sidwell doesn't have scientifically peer-reviewed numbers to back up a claim by an observer (such as myself) that the ranch is having a positive influence on *you-know-what*, but he isn't asking to be compensated by a carbon marketplace, either, unless you consider his grass-fed beef to be a type of carbon sequestration strategy (as I do). What if he did, however? Could a carbon marketplace be talked into considering mitigation as an "added value" to the beef produced from a ranch? Would customers be willing to pay? Would the Sidwells even want to go there? I think they might, though we didn't discuss it.

It was something to consider as I drove home across the parched landscape.

2

ABUNDANCE

On a fine August day, I flew to New England in search of abundance, but all I saw initially was devastation.

I arrived twenty-four hours after the region had been pummeled by Hurricane Irene, and the land looked like it had been stomped on by giants—entire corn fields had been flattened, trees were broken at crazy angles, and ominous-looking debris gathered along roadsides and at stream crossings. Adding to the unsettling imagery was a bright blue sky overhead and heavy traffic on the highway as I drove north, suggesting that all was well and normal—except for the mysterious roving band of giants out there someplace.

I was on the road to visit Dorn Cox, a young farmer who lives and works on his family's 250-acre organic farm, called Tuckaway, near Lee, New Hampshire. Dorn calls himself a "carbon farmer," meaning he thinks about carbon in everything he does. Confronting agriculture's addiction to hydrocarbons, for example, Tuckaway produces a significant amount of the energy it needs on-farm. Dorn does it with biodiesel—canola, specifically—which he and his family

grow on only 10 percent of the farm's land. This was big ne
thought a visit would be worthwhile.

I had learned about Dorn from a network of young farmers called the Greenhorns, which bills itself as a nontraditional grassroots organization dedicated to recruiting and supporting a new generation of farmers. They do this via social media, videos, podcasts, art projects, and activism, of course, but they also engage in old-fashioned networking—including mixers, barn dances, and an on-farm singles event called "weed dating."

The young farmers in the Greenhorns are nontraditional practitioners as well, in the sense that while their tomatoes, lettuce, broccoli, dairy products, and meat *look* the same as the produce from farms run by their parents' generation, the methods by which they achieve their harvests and the goals that motivate their work are very different. It's called "beyond organic," and it includes new ideas about no-till seeding, cover cropping, animal power, bioenergy production, carbon sequestration, open-source networking, and entrepreneurial business models aimed at creating local food systems. If that weren't enough, these young people are also defined by the substantial challenges they face, including the difficulty in finding affordable farmland, making a living in a world dominated by corporate agriculture, navigating a dense labyrinth of regulations, and adapting to the intensifying effects of climate change.

Such as roving bands of giants spawned by hurricanes.

I followed the path of destruction across the region. Almost six years to the day after Hurricane Katrina devastated New Orleans, Irene drenched New England with as much as 13 inches of rain, turning mild-mannered streams into raging torrents that knocked homes off their foundations, toppled rows of telephone poles over like matchsticks, and took huge bites out of roads. Dozens of rivers, primed by a rainy summer that had saturated soils, reached record flood levels. Vermont governor Peter Shumlin called it the worst flooding in a century. In upstate New York, winds flattened trees and toppled telephone poles like matchsticks. "It looks like somebody set a bomb off," said one resident. By the time Irene dissipated its energies over northern Maine, it had carved a trail of destruction from Florida to Canada. At least fifty-five deaths were blamed on the storm, and seven million homes and businesses lost electrical power. Extensive damage occurred along coastlines as a result of storm surge. Tornadoes caused significant property damage. The list went on and on. Ultimately, the price tag for Irene's damage would reach $15 billion, making it the sixth costliest hurricane in US history.

Had the giants spared Tuckaway? Apparently they had, because when I met Dorn a few days later, he was standing in a hayfield behind a home belonging to a University of New Hampshire professor, spreading wood ash carefully among a grid of study plots. He gave me a wave as I parked the car, putting the ash can on the ground. Farmer-thin, wearing muddy jeans, a yellow shirt, and a floppy straw hat that shaded intense blue eyes, Dorn extended a hand and gave me an energetic grin. "Everything alright back at the farm?" I asked. He reassured me that all was well—lots of rain, but no damage. "What's going on here?" I asked, nodding at the gridded plots, though I knew it was part of his Ph.D. research. "Just trying to figure out the best way to turn a hayfield into a farm without tilling it," he replied. "And create a food and energy system that puts more carbon into the soil than comes out." Was the professor okay with this? I asked. He's fine with it, Dorn reassured me. "There are a lot of these little fields behind people's houses. With some work they could be growing a great deal of produce," he said. "We just need to figure out a way to do it without using a plow."

As we walked across his study plots, Dorn explained his thinking. Conventionally, a modern farm requires a tractor and a plow in order to turn over the soil and furrow the land in preparation for seeding and fertilizing. In contrast, a no-till approach means a farmer can plant the seed directly into the soil, usually with a mechanical drill pulled behind a tractor or a horse. A thin slice is made in the soil by the drill as it moves along, but nothing resembling a furrow. The soil is not turned over, and whatever is growing on the surface is largely left intact. In fact, many no-till farmers plant a cover crop in the fall so that the soil will be kept cool, moist, and protected from the elements as the cash crop emerges from the ground in the spring or early summer. Dorn pointed at the hayfield as an example, indicating that the cover crop here was grass. He wants to know under what no-till conditions the cash crop—grains in this case—will grow best.

All of this was unconventional thinking, to put it mildly.

As a practice, plowing goes back at least five thousand years, evidenced by a famous hieroglyphic image of a plow on a wall in an ancient Egyptian temple. When agriculture first came into existence around ten thousand years ago, the handheld hoe was the main means of cultivating the soil, along with a digging stick for planting seeds. Four thousand years later, everything changed when oxen were domesticated in Mesopotamia and the Indus Valley, spurring the invention of the plow. This combination of animal power and human ingenuity has had a profound impact on the planet, allowing, for example, the global population to zoom from less than one million people to over seven billion today. Plowing was here to stay. As a technology, the plow has continued to be refined and adapted over the centuries in a textbook case of human creativity. As a science, however, plowing has had a rougher ride. Standing in the middle of his study plots, Dorn quoted the father of no-till agriculture, American agronomist and experimental farmer Edward Faulkner, who once said, "The truth is that nobody has ever exposed a scientific reason to till."

In 1943, Faulkner attempted to rock the farming establishment with his book *Plowman's Folly*, which challenged the orthodox view that the plow was as necessary to food production as breathing air is to life. He believed the exact opposite to be true. The plow, he argued from his experience, was an enemy of farmers. Loosening the soil by moldboard plow led to erosion, turning the soil over killed essential soil microbes by exposing them to heat and light, and the destruction of crop residues on the soil surface robbed plants of critical nutrients, organic matter, and shade. The answer, he insisted, was to toss the plow away. The response from the farming community was a deafening silence. Faulkner shouted into the void. He was considered mad by some and dismissed by the rest. Slowly, however, his radical idea took hold, especially as scientists backed up his observations with data on the alarming rates of soil erosion in farm fields. In the 1960s, his idea of no-till agriculture gained a foothold among a new generation of farmers focused on organic (nature-based) food production. By the 1980s it earned further support, as microbiologists explored the mysterious universe of life in the soil, quantifying Faulkner's hunch about the damage done to microbes by plowing. Meanwhile, no-till rose in popularity among hard-nosed corn and soybean farmers in the Midwest as they discovered its cost advantages over conventional plowing. In the end, profits and protozoa proved that Faulkner wasn't so mad after all.

Both profits and microbial soil life were very much on Dorn's mind there in the hayfield behind the professor's house, where he is attempting to combine his knowledge of organic farming with his training in high finance. I knew that Dorn had left Tuckaway after college for a job on Wall Street and then moved on to a private company in the high-tech sector. What I didn't know was that, like a good businessman, Dorn is trying to increase the return on his investment in the hayfield—the investment in this case being *carbon,* in the form of wood ash. Over the decades, carbon had drained away from New Hampshire's soils, largely as a result of plowing and erosion, and Dorn is trying to figure out what amounts are necessary, and in what proportion to other elements (such as nitrogen), to revitalize the soil's fertility once again.

"The soil here is like a bank to which I'm making a deposit of carbon that will create a natural form of compound interest," he explained. "Invest one seed, get one hundred back, return the carbon residue to the soil, and invest seed once again next season, and get one hundred twenty back. This absolute return is the real discount rate, and the carbon the real collateral. Any economic returns achieved above the real biological rate of return are by definition extractive and, therefore, unearned."

And it's *earned* income that Dorn is after; he calls it the basis for *real wealth.*

The true costs of conventional agriculture, which must include a full accounting of fossil fuel extraction and pollution, Dorn said, far exceed the biological rate of return. These costs deplete bank accounts by removing their carbon deposits. No-till, in contrast, builds compound interest and creates real wealth because it builds life in the soil. Life creates life, and with it, abundance. And abundance is what Dorn is after ultimately, not simply for himself and his family, but for his community, his home state, and the nation as well.

In America, as in most nations, economic theory and practice is dominated by *scarcity thinking*, which is the belief that there's not enough of something to go around. Oil is a classic example. As oil becomes scarcer and more difficult to extract from underground, it becomes more valuable, and thus more profitable to those who supply it—and more expensive to those who need it. This creates an important social impact to go along with the economic one. When a commodity becomes scarce, we as a society start thinking about it obsessively—Where is it? How do we get at it? Why does it cost so damn much?—instead of investigating more abundant alternatives, such as solar energy. Psychologically, scarcity thinking is fear-based; it compels us to do things like hoard, compete, fight—and act greedily, selfishly, and dishonestly. It creates winners and losers. In contrast, *abundance thinking* is the belief that there's plenty for everyone. Soil is a classic example. There's a lot of good, rich soil in the nation, Dorn pointed out. It could be doing so much more for us if we would only look at it through the lens of abundance, not scarcity.

This is by no means a new idea, Dorn explained. In the 1700s, a group of French economists who called themselves *physiocrats* ("government by nature") argued that all wealth originates from the land, making farming the only truly productive enterprise. All other work was seen as extractive or transformational of the original value created by farmers. All agricultural products, they believed, circulated through an economy like blood through a body and were just as essential for well-being and long life. Physiocratic thought influenced Thomas Jefferson, Benjamin Franklin, and Adam Smith—whose seminal work, *The Wealth of Nations*, reads like an agrarian manifesto, Dorn said. However, physiocracy failed to take hold, mostly because the bounty of natural resources newly discovered in Asia, Africa, and the Americas appeared to be unlimited. Careful stewardship of the land took a distant backseat to rapid and dramatic resource exploitation, leading eventually to scarcity anxiety as supposedly bottomless wells of resources began to run dry.

It's only now, Dorn believes, as we bump up against significant and unbending environmental limits, that the advantages of a physiocratic-style economy are becoming evident again. By employing the lens of abundance thinking, we can suddenly see the world as bountiful and hopeful. Take Dorn's home.

Once upon a time, New Hampshire grew much of its own food. In the 1830s, Dorn said, two towns raised more sheep than are raised in all of New England today, and for many decades New Hampshire farmers grew thousands of acres of wheat, more than enough to feed its citizens. Unfortunately, shortsighted management created a legacy of overgrazing and overlogging in the state, resulting in depleted soils and eroded land—a story common throughout the region (and elsewhere). Over time, nearly all the grain and dairy farmers trickled away to greener pastures, and New Hampshire's ability to feed and heat itself steadily declined. The collapse of the state's industrial economy in the late nineteenth century led to a general exodus of population, a trend reversed only recently as high-tech companies, telecommuters, and wealthy second-home owners began to move in. Today, only 5 percent of New Hampshire is farmland, which means agriculture is essentially a cottage industry.

"New Hampshire is the 'Live Free or Die' state, known for the independent spirit of its citizens," Dorn said. "But despite this heritage, it is now one of the most dependent states in the union, relying almost wholly on imported food and fuel."

New Hampshire has a population of 1.3 million people. If only 13,000 of them (1 percent) became new farmers the state could feed itself, Dorn said. This is possible because New Hampshire has (1) lots of rain and snow, (2) good agricultural soils, (3) plenty of market potential, (4) a strong educational system, and (5) wealth—i.e., capital—which is necessary to invest in new food systems. In other words, the state has an abundance of possibility. What it lacks, he said, is knowledge and a willingness to change its ways of thinking. Over 40 percent of New Hampshire's soils are rich enough to be producing food, and yet only a tiny fraction of the population is engaged in farming. It's the same situation with fuel. The majority of homes in the state are heated with oil, Dorn told me, and yet two of the most common complaints he hears are about the high cost of oil and the low price for wood—in one of the most heavily forested states in the nation.

Scarcity in a land of plenty.

"It's a cultural paradox," said Dorn. "With lots of fertile soil, forests, water, and capable people, why can't we make an independent, abundant living once more?"

This is a question Dorn asked himself ten years ago when he returned to his family's farm after a personal exodus that took him first to Wall Street and then into corporate work in Buenos Aires, Hong Kong, and other exotic locations around the planet. He left home thinking about scarcity and came back thinking about abundance—a voyage of discovery that says a lot about our times and Dorn's generation.

Growing up, Dorn had heard repeatedly from friends and neighbors that farming was something to get away from, so after studying international agriculture and rural development at Cornell University, Dorn moved to New York City, where he worked in finance as a certified securities broker (in a suit and tie), in an office overlooking the Hudson River. The future of family farming was bleak for a young person, he thought. Profits, security, and personal growth were scarce, which is why the siren call of Wall Street was hard to resist. However, after twelve months the song turned sour, so Dorn quit his job, joined a technology company, and moved to Buenos Aires—where he began to think hard about renewable energy, a topic that had intrigued him from a young age. As he traveled for the software firm, burning lots of jet fuel, Dorn turned over in his mind the challenge of replacing fossil fuels with biologically renewable sources, such as the biofuels being developed in Brazil from its abundant sugarcane fields. Making the switch wasn't simply about technology, Dorn suspected. It had more to do with how we *thought*, the way we looked at the world, the things we chose to see as valuable and what we chose to ignore as "uneconomic," such as sunlight and soil. And the more he thought about all these things, the more Dorn wanted to go home. While on a hydrofoil on a business trip to Macau he wrote this note to himself: "I deeply miss my farm and my family."

"It turned out that the farm had never stopped being part of me," Dorn explained, "so I left the cities, met my wife, and returned to the land where four generations of my family now live and work."

His exodus had come full circle.

Leaving the professor's hayfield, I followed Dorn in my car down a leafy road near the university. We made a sharp turn onto a narrow lane, past a small sign posted on a tree that said simply TUCKAWAY FARM. Dorn's parents, Charles and Laurel, bought the property in the 1970s as part of their generation's back-to-the-land movement, with the radical goal of growing veggies, blueberries, and hay—radical because they were the first farmers in the area to go organic. The farm had been in agriculture for many decades, but it was on a downward slide as it lost soil fertility due to relentless plowing and annual applications of chemical fertilizers, herbicides, and pesticides. By going organic—that is, by ceasing every agricultural practice that ended with the killing suffix "-cide"—Dorn's parents checked the downward trend in land health. Under their stewardship, the land was given a chance to recover and grow again as the microbial life in the soil bounced back. The land's fertility stabilized and life was good. Dorn, however, had come back to the farm with new ideas, including a nascent theory of abundance that he wanted to try out. At heart was a guiding question: Could the fertility of the land be restored to what it was long ago using only regenerative, natural processes?

And a second one: Could New Hampshire also be restored so that it could feed itself again?

Dorn decided to test these questions by focusing first on energy. For decades, his parents had relied on fossil fuels to power their tractors and other equipment, but Dorn decided to convert as much of the farm's energy use to biodiesel as possible. He knew that the process of producing biofuels could be scaled to a single farm, which meant that if he could find the right feedstock they would be able to bypass big parts of the petroleum supply chain. His first thought involved sunflowers, which Dorn had observed growing under similar conditions in Argentina. In 2002, after getting a "thumbs-up" from experts at the University of New Hampshire, sunflowers were planted at Tuckaway and an oil press was purchased. When the harvest came in, it produced 60 to 80 gallons of oil per acre, Dorn said, as well as a 30 percent protein meal as a by-product of the pressing, which could be used as feed for the animals. This was exciting news.

It also meant that small cracks in New Hampshire's cultural paradox could potentially be widened into bigger ones.

After introductions to his family, Dorn and I walked a short distance down the road to what looked to me like a Coca-Cola truck. It *was* a Coca-Cola truck—a used one. As we reached it, Dorn pushed up a panel in a section, revealing not boxes of sodas, but a large container with various hoses coming and going. It was a part of the biodiesel production process, Dorn said as he pushed up other panels. Each big step in production had its own compartment, which not only made things tidy but meant the process could be easily replicated by other farmers (provided they had their own Coca-Cola truck). Even though making biodiesel involves what Dorn called "off-the-shelf technology," it took some time and a lot of tinkering before Dorn and a few buddies, through a nonprofit called Green Start, created a biodiesel prototype to their satisfaction. It can make 200 gallons of fuel in an afternoon, Dorn said, depending on the feedstock. And housed in a Coca-Cola truck, it can travel from farm to farm, to teach or make fuel. As for the sunflowers, Dorn eventually added canola, an oilseed plant developed in Canada from rapeseed, an ancient source of oil for food.

Dorn explained that there were other benefits to biodiesel besides its ability to be produced on the farm: it can be used in a conventional diesel engine very easily; it is cleaner burning than petroleum-based diesel; it's a natural lubricant, which makes engine parts last longer; it doesn't pose a threat to human health; it's 100 percent biodegradable; it is safe to store, transport, and clean up; and, critically, it has a positive energy balance, meaning for every unit of energy used to make a gallon of biodiesel, as many as three units of energy are gained. This is important to Dorn because his goal for the entire farm is a *net carbon energy balance*—that is, the farm creates more energy (as output) than it consumes (as input). It's the real wealth thing again: he's earning interest from his investment, not depleting accounts. Creating a net energy balance is a tall order, of course, but Dorn thinks it is achievable—no, it *must be* achievable if we're going to live sustainably on this planet for the long run.

Biodiesel isn't the only path to this goal, Dorn said, which is why he led me next to the barn, instead of heading out to the fields. He wanted me to see the horses. Dorn's sister and brother-in-law share his concern about energy and sustainability, but they don't like anything that puffs diesel smoke, whether bio-based or not. That's why they turned to a power source nearly as old as agriculture: draft horses. In fact, they've become just as passionate about old-fashioned horse power as Dorn is about biodiesel, and their respective passions sometimes become spirited debates at the supper table. It's all good, Dorn said with a smile: same goal, different methods. Unfortunately, the horses weren't in the barn. Apparently, the team was out working on the farm someplace, likely under the stewardship of Dorn's dad, who, Dorn said, has become smitten with the powerful animals. I ask Dorn how his parents feel about all these newfangled ideas and practices that their children have brought back to the farm. "They're intrigued," he said, with a smile that suggested he and his sister aren't the only ones engaged in spirited debates over supper.

On the walk to the farm field, Dorn said Tuckaway is a member of a community-supported agriculture (CSA) program and recently added biodiesel to its diverse produce, which includes grains, berries, hay, timber, firewood, maple syrup, meat, and vegetables. For the first two seasons he grew the biodiesel-destined sunflowers the old-fashioned way—with a plow. It wasn't until he analyzed the energy costs associated with dragging a heavy iron plow across New Hampshire's rocky soil that he realized the costs of tilling were too high from an energy balance perspective. Dorn decided to look around for an alternative. He knew about no-till farming, but he also knew that one of its disadvantages was its lack of weed control. If the ground isn't turned over with a plow, the weeds say "thank you very much" for all that undisturbed soil and start growing vigorously, sometimes elbowing out the cash crop. To check weeds in a no-till system, many farmers apply synthetic herbicides to their fields. They'll also spray pesticides to keep the bugs in check. Additionally, many conventional no-till farmers will use genetically modified seeds, often in combination with chemical herbicides. All of this is verboten in an organic farming system, of course. As a certified organic farm, Tuckaway found itself between a rock and a hard plow, so to speak.

Fortunately, Dorn discovered an answer to his dilemma at the Rodale Institute, located north of Philadelphia. Named for J. I. Rodale, the founder of organic agriculture in America, the institute has been a leader in research and education in organic farming systems since 1948. That's the year when Rodale penned his famous equation: Healthy Soil = Healthy Food = Healthy People. This declaration might seem clichéd now, but in its day it was as radical as Edward Faulkner's demand that we ditch the plow and was received by farmers just as enthusiastically. The chemical fertilizer, pesticide, and insecticide industry was about to take off in 1948—transforming our soil, food, and health in very profound ways—and it steadfastly refused to hear Rodale's message. So, Rodale did the only sensible things open to him at the time: start a publishing business, found a research institute, and wait. Sooner or later, he surmised correctly, organic farming would catch on.

What Dorn discovered at Rodale was a way to do organic no-till agriculture. The practice was developed by Jeff Moyer, the institute's longtime farm director, who came up with an innovative way to combine the two farming systems. It began as many good ideas do—by accident. One day, Moyer noticed that as he drove in and out of a farm field on his tractor, the wheels had crushed and killed a plant called hairy vetch, which grew along the field's edges. Vetch is a winter-tolerant, nitrogen-fixing legume that organic farmers often plant as a cover crop in their fields. Seeing that the vetch was still alive where he had not driven over it, Moyer realized he had "crimped" the plants with the tractor's wheels, causing them to die *without* causing them to detach from the soil, as cutting or harvesting would do. This intrigued Moyer because, by remaining attached to the soil, the dead vetch could become a type of in situ mulch for the soil. Normally, cover crops are harvested, composted, and returned later to the field as mulch. Moyer's accidental discovery changed this equation dramatically: he could now crimp the cover crop instead!

However, no mechanical piece of equipment existed to do this job specifically, so Moyer decided to invent one. After a great deal of trial and error, he and a colleague settled on a design for what they call a "roller-crimper"—a hollow metal cylinder to which shallow metal ribs have been welded in a chevron design (like tractor tires). The roller-crimper is mounted in front of a tractor or behind a horse, and as it rolls along through a field it crimps the cover crop, breaking the plant stalks and killing them. The weight of the crimper can be adjusted by either adding water to the cylinder or removing it in order to achieve the desired effect. In other words, the roller-crimper merged no-till with organic farming: no synthetic "-cides," no transgenic seeds, and no plow. Voilà!

As developed by Moyer and colleagues, there are four basic steps to organic no-till:

1. Protect the soil and keep down the weeds in a farm field by planting a winter-hardy cover crop in the fall, such as vetch, barley, wheat, rye, or oats.
2. When the cover crop reaches maturity in the spring, the farmer knocks it down with a roller-crimper.
3. The farmer plants a cash crop into the crimped cover crop with a no-till drill, usually at the same time as crimping (crimper in front of the tractor, drill pulled behind), and then the cash crop grows up through the crimped cover crop.
4. After harvest in the fall, the organic residue of both crops can be disked into the soil, if the farmer wants, as next year's cover crop is planted. All together, the use of a cover crop and a roller-crimper creates a dense mat of organic material on the soil surface that smothers weeds while providing nutrients, shade, and moisture to the cash crop.

Dorn jumped at the idea. He thought it literally rolled the best ideas in agriculture together: organic production, no-till, and a positive energy balance (this method requires only two passes with a tractor in contrast to the ten or twelve needed in a conventional system). He put it to work on Tuckaway as soon as he could.

As we walked to the edge of a field that Dorn had cover-cropped and crimped the previous spring, he listed the various benefits of the Rodale process: soil is built by the decomposing cover crop; erosion is reduced substantially; nearly all annual weeds are smothered; cover crop roots increase nutrient cycling in the soil, including carbon and nitrogen; carbon dioxide is sequestered in the soil; greenhouse gas emissions are reduced; costs are low; and the roller-crimper is easy to use and maintain. To this list, Dorn added his own contribution: a positive energy balance, since he ran the tractor with farm-made biodiesel. Taken all together, it was a big step toward answering his question about restoring the fertility of the land using only regenerative processes.

"Organic no-till," he said, "means you can eliminate three things from your to-do list: increase biodiversity, reduce erosion, and manage organic matter in the field. To a farm like Tuckaway, these are huge."

There are downsides, however, as Dorn noted: cover crops are extra work and add extra costs; they require water, sometimes a lot of it (which can make the practice problematic in arid environments); perennial weeds can still be a nuisance, especially if the cover crop is thin or didn't "take" properly across the farm field; choosing the correct cover crop for your land and matching it to the needs of the cash crop can be tricky, requiring experimentation; rolling the crimper too early in the season can be a costly mistake because if the cover crop doesn't die completely it will compete with the cash crop for water and nutrients; and, like anything new, success requires a great deal of patience. For Dorn, however, the positives far outweigh the negatives, which is why he calls organic no-till the "holy grail" of organic farming, crediting Moyer and Rodale for a major development.

It got better.

As Dorn explored this new method, he made an unexpected discovery: he was building abundant life in the soil. Research at Rodale and other places showed that traditional plowing practices destroy the microbial universe underground, mostly by exposing beneficial protozoa, nematodes, fungi, and other forms of life to the killing effects of sunlight, wind, and heat. The plow itself also tore delicate fungi to pieces. These micro-critters are the key to soil fertility, which is why synthetic fertilizers are required in conventional systems—to replace the fertility lost by their mass slaughter. Dorn also learned that plowing releases large amounts of stored carbon into the atmosphere, adding to the planet's greenhouse gas problem. When soil is turned over, the sudden access to oxygen speeds up the biological decomposition process, by which microbes (before they die) eat up organic matter and "burp" carbon dioxide into the air. Repeated plowing eventually depletes the soil of its carbon stocks. Plowing also encourages erosion, which means carbon literally washes or blows away. Lastly, synthetic fertilizers burn up carbon chemically, releasing it into the air.

All these facts came as revelations to Dorn, who had not focused on the role that micro-critters, carbon, and other elements played in the health of Tuckaway's soil.

"I realized that everything that we grew on the surface was peripheral to preserving and building healthy soil," he explained. "In order to remain viable into the next century and beyond, a new, soil-centric approach had to emerge here. I saw that we had to deposit more carbon than we took out, so the soil will be more resilient and provide a more regular return on our investment, no matter what the crop."

This led to another realization: to achieve his goal of true abundance, Dorn was going to need help.

"It got overwhelming pretty quickly," he said with a smile. "There was clearly too much trial and error for one farm to bear. It became important to find other farmers willing to share the risk in developing new growing techniques."

This realization first dawned when Dorn decided to add grains into Tuckaway's crop rotation. Asking around, he discovered that no one in New Hampshire had actually grown grains organically since the Civil War! Fortunately, Dorn found help from the US Department of Agriculture as well as a few growers scattered in other states. Eventually, he and several local farmers organized the

Great Bay Grain Cooperative with the goal of sharing knowledge and working collaboratively to build a local grain market via direct sales to customers. Following Dorn's lead, many of the members in the cooperative also aimed to produce as much of their inputs on-farm as possible, such as biodiesel, rather than buying retail. They also share their experiences, problem-solve together, and help at harvesttime. All of these things, Dorn told me, have contributed to the cooperative's success: the original five farms have doubled to ten farms in four years, sales are strong, and the future is bright.

After a 150-year absence, organic grains had returned to New Hampshire.

This was just the start, however. To demonstrate what happened next, Dorn and I left the fields behind, crossed the farm, and climbed into my rental car. Dorn wanted me to visit two neighboring farms, both run by Greenhorns. As we drove, Dorn explained that "getting help" these days, especially among young farmers, meant more than cooperatives and mutual assistance at harvesttime. It also meant the Internet—specifically a philosophy called *open-source networking*. For Dorn, this network has become as important to the success of Tuckaway and his theory of abundance as carbon management in the soil.

In his old job for the high-tech company, Dorn explained, he became a specialist in systems analysis, which he used to build networked enterprises based on open-source software—meaning software that was freely available to anyone who had access to the Internet. This software allowed him to manage complexity—orders, contracts, projects in diverse locations around the planet—as well as teach others how to manage similar amounts of complexity in their specific business environment. The key was information sharing and computing power. Dorn compares it to trying to understand a specific DNA sequence in a plant or animal. No single person can comprehend an entire sequence, but with lots of networked computers and data sharing it's possible to see the whole DNA picture. It's how complex systems can be managed—such as running a 250-acre organic farm.

Thanks to technology, farmers today have an unprecedented ability to document, analyze, and understand what's happening on their land. This knowledge is then made available to other farmers through the "creative commons" provided by open-source software, Dorn said, rather than being buried in academic journals, locked up in patents and licenses, or tucked away in a corner of a farmer's brain. Dorn likens it to the technology used by doctors to diagnosis a patient's complaint and prescribe an effective course of action. It's the same on a farm, though instead of curing an illness Dorn employs sophisticated monitoring and feedback systems to increase the land's "biological velocity," as he puts it, in order to create abundance and real wealth.

As we drove through a large town, Dorn quoted Adam Smith from *The Wealth of Nations*: "To improve land with profit, like all other commercial projects, requires an exact attention to small savings and small gains." On a farm these small savings include soil moisture levels, nutrient indicators, microbial activity, carbon content, and plant and animal interactions. However, for most of farming's history, the ability to observe these details has been limited to what we can detect with our senses—sight, smell, touch, taste—and how our brain, via experience, interprets this narrow range of data. Modern technology has expanded this range exponentially, and thanks to the Internet the data is democratically available to anyone who seeks it, without the mitigation of an expert, academic researcher, or paid consultant.

As we pulled away from the town, Dorn explained that he had tested his open-source philosophy on the biodiesel Coca-Cola truck he and his friends constructed. Everything they did, all the designs and adjustments, were made freely available via the Web. Nothing was proprietary. No one filed a patent, no one hoarded secret formulas, no one hired a lawyer to sue a competitor or demand compensation. The give-and-take that resulted helped them build a better biodiesel processor. It's been the same with the grain cooperative and even his Ph.D. research, Dorn said. Ditto with his farming peers. Free-flowing dialogue and unobstructed access to knowledge, innovation, and data are keystones to the young farmers movement today—as are the advanced technology and social media they regularly employ (naturally).

For example, Dorn said he uses an online network called Farm Hack nearly every day to answer questions, brainstorm, share ideas, or discuss needed changes to farm policy. An initiative of the National Young Farmers Coalition, Farm Hack combines online Q&A-style problem solving with an active network that can help organize real-life conferences and meet-ups on farms. The main goal of Farm Hack is for users to learn from one another, Dorn said, and it is especially valuable for anyone who was not raised in a farming environment. Many problems that young farmers face have been resolved before, but these answers can be difficult to find. Farm Hack enables farmers to find inexpensive and scaled-to-fit solutions. The meet-ups are usually held on a campus and involve engineers, electricians, carpenters, and many others in addition to farmers. Dorn, in fact, organized the second-ever Farm Hack meet-up, a gathering that involved four institutions and drew participants from six states.

"They are also a lot of fun. There's great food, conversation, and a lot of time out in the fabrication shop making things," he said with a smile.

Dorn also regularly checks an online forum called Public Lab, which develops and shares ultra-low-cost technology. When Dorn needed an overhead view of his test plots in the professor's hayfield, he went to Public Lab and found an answer: a $100 balloon-mounted camera that floats 25 feet above the ground! The infrared images taken by the camera have yielded important data, he said, that would otherwise have taken weeks of labor to collect. Later, thanks to the Internet again, he discovered a way to document larger amounts of land using a camera mounted in a remote-controlled aircraft, like the kind you might have flown as a kid on a soccer field. Its cost: $200.

Dorn believes that online forums such as Farm Hack and Public Lab, along with traditional cooperatives and collaborative research projects with other farms, are as important to modern farming today as walking the fields each day.

"The complexity of my farming operation would be unmanageable without them," he said. "I'm certain that open-source knowledge sharing will revolutionize agriculture just as Wikipedia has revolutionized the encyclopedia."

Arriving at our destination, we pulled into a small parking lot in front of a single-room farm store that bustled with young people. As Dorn headed inside to arrange introductions, I climbed out of the car, stood, and stretched my arms high into the humid air. Nearby, a very large and ancient-looking tree towered over the lot, casting an inviting amount of shade on what felt like an oppressively muggy day to this drylander. I'm used to air that crackles. I'm also used to trees that are not much taller than I am. As I admired the venerable tree, a young man suddenly glided to its base on the most dilapidated bicycle I had ever seen. He wore a T-shirt, overalls, and a youthful beard. As he hopped off the bike, he gave me a warm smile and a nod of his head. After I nodded back, he pulled a smart phone from a dungarees pocket with a practiced motion and studied its face as he walked toward the store. In a minute, he disappeared into the energetic swirl at the door.

I drifted over to the towering tree, needing a brief respite from the world's unceasing busy-ness. Maybe I could stand sentinel for a while, I thought, and keep a lookout for those roving bands of giants. Lacking a smart phone, however, I'd have to do it the old-fashioned way—with my faulty senses. I scanned the road. The coast was clear. My thoughts turned to everything I had heard today, to carbon inputs, energy balances, computing power, paradigm shifts, and Dorn's ideas about abundance. I admitted to myself that I was a bit bewildered by it all. "Organic farming isn't rocket science," I thought, "it's much more complex." It's much more exciting as well, and much more necessary, especially the type of farming Dorn and his fellow Greenhorns are trying to establish. As for the complexity of it all, well, as the young man's practiced fetching of his phone indicated, managing complexity is second nature to this generation, thankfully.

I also considered his smile. It wasn't just the technology that mattered; success still depends on relationships and on community—and not just the virtual kind. Barn dances, weed dating, meet-ups, and smiles were necessary to put ideas of abundance into practice. Dorn needed a lot of data and computers to manage his farm's complexity, but he also needed friends and laughter and handshakes too, especially if they were going to be successful in warding off giants.

I scanned the road once more, turned, and headed for the swirl at the door.

What Dorn discovered is the *soil food web*, a concept used by ecologists to describe "what eats what" in a given spot on the planet. A food web sometimes called a "food chain" or "food cycle," but in all cases it refers to the flow of energy through a system created by the process of eating, digesting, defecating, and decomposing. Round and round. Birds and bees do it. Lions and vultures do it. We do it. So do protozoa and nematodes. The soil food web describes the complex community of life belowground and the mind-boggling variety of relationships between plant roots, micro-critters, minerals, water, and energy—a web of activity that sustains *all* terrestrial life on Earth. Every living thing aboveground—whether it's hairy vetch, a mighty ponderosa pine, an elusive coyote, an endangered bird, a busy beaver, or an equally busy human being—depends on the vast universe of life in the soil, and does so for a simple reason: we must eat. Every living thing needs energy, and most of this energy comes from food. No food, no life. And where does this energy originate? In the soil. If we are what we eat, as the saying goes, then we'd better pay attention to the universe below our feet.

And at the heart of this micro-world is carbon, the glue that holds the web together.

Think of what happens belowground as a type of biological engine. Plants convert sunlight and carbon dioxide into jet fuel (liquid carbon), which is then sent to the factory underground via an elaborate root system to nourish the micro-critters at work on the engine. The job of these critters is to break down organic matter into smaller and smaller bits, "dig up" minerals from the soil to supply the engine, and facilitate deliveries of fresh water. To do this they need a steady supply of oxygen, which is provided by tiny air pockets in the soil. They also need a healthy work environment, free from toxic chemicals, thuggish gangs of viruses, and cataclysmic events such as plowing. It's also good to have large friends, such as earthworms, to help with the labor. And like all workers, the critters need to rest, eat a good diet, and pass gas—carbon dioxide in this case, which makes it way up to the surface and out into the atmosphere. If all the parts of the biological engine are healthy and working together efficiently, then the bountiful end product is *humus*—good, rich, dark organic soil. What kind of humus gets produced depends on what raw materials are available to the engine, including soil type, water quantity, microbial activity, root depth, and the amount of organic matter lying around. The thickness of the humus influences the productivity of the engine, including how much 'exhaust' it emits (as a greenhouse gas).

As the engine grows, all of its parts grow with it: plant roots get bigger and extend deeper into the soil, sending larger amounts of jet fuel to the micro-workers, who grow more numerous, excavate more minerals, store more carbon, and pass more gas. As the humus expands, all this work makes it "fluffier"—more porous with air pockets—and in turn it holds more water. This encourages additional biological activity, which creates more humus, round and round. Life begetting more life, turning *dirt* into *soil*. Dirt is chemistry: individual particles, minerals, and elements—nitrogen, calcium, phosphorus, potassium. Soil is biology: bacteria, fungi, protozoa, nematodes, earthworms, reproduction, growth, life. Getting food to grow in dirt is chiefly a matter of getting a chemical formula right and applying it mechanically according to a calculated prescription. Getting food to grow in soil, in contrast, is a matter of getting the biology right.

Turning dirt into soil isn't difficult, though it can be time-consuming depending on your goals. Turning soil into dirt, however, is quick and easy. Just follow the recipe: Plow. Add chemicals. Mix well. Repeat.

The key element in revving up the engine of a soil food web, after sunlight, is carbon. To explain, let's follow three carbon molecules from the air to three separate destinations. Imagine these molecules enjoying life as a gas, each accompanied by two oxygen molecules, as they joyride side by side through the air without a care in the world. Suddenly, all three smack into a green, leafy something and quickly pass from the bright light of the atmosphere into the dark, vascular world of a plant. Now their carefree existence turns into a wild toboggan ride of photosynthesis. The three molecules go through a series of transformative twists, turns, and drops as they travel through the plant, bathed in green, drenched in water, stripped of their oxygen buddies, and eventually picking up new molecular passengers, including hydrogen, nitrogen, and more carbon. At the end of their wild ride, the molecules are no longer part of a gas, having become instead part of a sugary carbohydrate called *glucose*, a vital source of energy for the plant. At this point a new ride begins and our three carbon molecules are quickly sent in three separate directions.

The first molecule concludes its journey in a leaf cell, where the glucose is converted by the plant into a kind of biological battery called *starch*, which it stores for later use, such as in winter, when photosynthesis is turned off. (Other uses of glucose by the plant include respiration, creating the sweetness in fruit, conversion into cellulose for cell-wall strengthening, conversion into fatty lipids for storage in seeds, and conversion into proteins, which are an important source of food for all living things.) Inside the leaf, our carbon molecule rests quietly in its cell, waiting patiently to be summoned, when the leaf is suddenly ripped from its host by a hungry herbivore. After a brief but tumultuous ride through grinding teeth, the molecule slides downward into a smelly stomach and eventually passes into the animal's digestive tract, where the starch is processed and the carbon absorbed into a cell of muscle tissue. A month later, the cycle is completed when the animal breaks a leg and dies in the wild. As it decomposes, the carbon molecule is eventually exposed to the air, where it picks up two swinging oxygen atoms and rises upward to begin the joyride all over again.

The second carbon molecule shoots belowground through the plant's stem and then slows to a crawl as it reaches the tip of a slender root not far below the surface of the ground. This is where the labile carbon is located, but before the molecule can join the party, the plant suddenly shudders as half of its leaves are wrenched away by the hungry herbivore. This causes the plant to send an emergency signal to its roots: *retreat!* To help recover its vigor and grow new leaves, the plant must now use the glucose stored in its root cellar, ordering the supplies upward. It's not a crisis, however (unless the hungry herbivore comes back around for a second bite), because by removing last year's dead grass along with this year's green growth, the herbivore has freed up the plant to grow unimpeded. One of the plant's first responses is to slough off the tips of its roots to conserve energy. That means our second carbon molecule finds itself detached and isolated from the rest of the plant, lost and lonely in the soil—but not for long.

Soon, the decaying root tips attract the attention of a host of hungry microbes and other critters, including protozoa, nematodes, fungi, earthworms, arthropods, and a huge variety of bacteria (there are more microbes in a teaspoon of soil than there are humans on the planet). Bacteria go to work first. These are single-celled creatures with one goal in mind: eat! They are particularly ravenous for carbon, and after they have digested a bunch of it they become attractive to predatory critters in the neighborhood. Soon, a feeding frenzy begins. Our particular carbon molecule disappears down the throat of a ubiquitous nematode, which is a type of tiny worm. There are approximately one million different species of nematodes on the planet, found in every type of ecosystem, accounting for 80 percent of *all creatures* in existence on Earth. That makes for a great deal of eating and pooping going on belowground. In this case, the nematode eventually excretes our molecule into the soil, where it bonds with two sly oxygen molecules hanging out (smoking cigarettes, no doubt) in the tiny air pockets and becomes part of a carbon dioxide molecule once again. Eventually this new CO_2 molecule makes its way up to the soil surface and back into the atmosphere to join zillions of its buddies for another wild ride. Round and round.

Meanwhile, our third carbon molecule has traveled to a deep root, below the labile level, where it ends up inside a fungus instead of a nematode—and not just any fungus, but one of the heroic mycorrhizal variety. These are long, skinny filaments that live on the surface of plant roots, with which they share a symbiotic relationship, trading essential nutrients and minerals back and forth (*mycorrhiza* is Greek for *fungus + root*, or "I'll scratch your carbon if you'll scratch mine"). The fungus-root mutualism reduces a plant's susceptibility to disease and increases its tolerance to adverse conditions, including prolonged drought spells or salty soils. Fungi in general are best known to humans as the source of mushrooms, yeasts, and the molds that make cheeses tasty, ruin houses in humid climates, and produce antibiotics. Like plants, animals, and bacteria, fungi form their own taxonomic kingdom. There are an estimated two to five million individual species of fungi on the planet, of which less than 5 percent have been formally classified by taxonomists. In the soil, fungi can be "good guys" or "bad guys" depending on your perspective (the "bad guys" can cause a variety of diseases). Our third carbon molecule ends up in a good guy called an arbuscular mycorrhizal fungus—a *really* good guy, as we'll see.

After absorbing the bit of glucose containing our molecule from the plant root, the skinny arbuscular fungus next pushes the carbon into one of its *hyphae*—hairlike projections that extend as much as 2 inches into the soil in a never-ending search for nutrients. Then, in a process that is not completely understood by scientists, the carbon molecule is extruded from the hyphae in a sticky protein, which coats it in a kind of gummy armor.

This protein is called *glomalin*—the unsung hero of Carbon Country.

Glomalin is one of nature's superglues. As a plant grows, mycorrhizal fungi move downward to colonize new roots, detaching old glomalin-coated hyphae as they go and growing new ones. The now free-floating glomalin quickly binds itself to loose sand, silt, and clay particles. Soon, small clumps of glomalin-glued particles form larger and larger aggregates, kind of like a vast, intricate tinker-toy construction. As the aggregates grow bigger they become stronger and more stable, making the soil increasingly resistant to wind and water erosion. This process also makes the soil more porous (fluffy), with lots of tiny pockets in between the tinker-toy aggregates, and this facilitates oxygen infiltration, water transport, micro-critter movement, and nutrient transfer. Next stop: humus. You can feel glomalin, by the way. It's what gives soil its *tilth*—the rich, smooth texture that tells experienced farmers and gardeners that they are holding great soil in their hands. To create tilth, the soil engine needs both biology and chemistry working together, and glomalin is the glue that binds them.

It gets better.

Glomalin itself is a tough protein. It can exist for up to fifty years without decaying or dissolving. When locked into the stable tinker-toy structure of humus, it can persistent for even longer periods of time. Healthy soils have a lot of glomalin, which means this: since glomalin is 30 to 40 percent carbon, it is the ideal safe-deposit box for the long-term sequestration of atmospheric carbon dioxide. This is what scientists call "deep carbon"—the kind that stays in the soil for decades, or longer. There are fewer hungry microbes this deep in the soil, which adds to the stability and longevity of the carbon storage. It's a simple equation: lots of deep glomalin = lots of secure carbon storage. It's also a fragile equation, however. A plow can destroy this safe-deposit box in a heartbeat, releasing its carboniferous contents back into the atmosphere. Plows also tear mycorrhizal fungi into bits, slaughtering them in droves, putting an end to our unsung heroes.

No one knew glomalin existed until it was discovered in 1996 by Sara Wright, a soil scientist with the US Department of Agriculture's Agricultural Research Service in Maryland. She named it after Glomales, the taxonomic order that includes arbuscular mycorrhizal fungi. Not only did she uncover its role in soil building and carbon sequestration, but a subsequent four-year research project under her direction demonstrated that levels of glomalin could be maintained *and raised* with regenerative farming practices, including no-till planting. In the study, Wright observed that glomalin levels rose each year after no-till was implemented, from 1.3 milligrams per gram of soil (mg/g) after the first year to 1.7 mg/g after the third. A control plot in a nearby field that was plowed and planted each year had only 0.7 mg/g. In a further comparison, the soil under a fifteen-year-old buffer strip of grass had 2.7 mg/g of glomalin. She also discovered that some plants don't attract arbuscular fungi to their roots, including broccoli, cabbage, cauliflower, mustards, rapeseed, and canola (perhaps to Dorn Cox's surprise).

Before 1996, determining the carbon content of a farm's soil was largely based on measuring its soil organic matter (SOM), which is roughly 58 percent carbon. Thanks to the discovery of glomalin, soil carbon can now be measured quite precisely. This sort of data is very useful in determining how much deep carbon a specific farming or ranching practice is sequestering.[1] It has economic implications as well, since carbon trading markets, such as the ones recently established in California and Australia, could potentially use levels of glomalin as a "currency" to pay landowners for mitigating carbon dioxide pollution. Employ a farming or ranching practice that is known scientifically to increase levels of glomalin, and get compensated financially!

Wright's work answered farmer Faulkner's complaint from decades ago: there is, in fact, a solid scientific reason *not to till* the soil. Plowing destroys the network of life underground, including the soil's capacity to biosequester carbon dioxide. Faulkner could only guess at the real reasons to go plowless; ditto with Rodale and his concern over synthetic pesticides and herbicides. Their objections were intuitive, based on their experience and their five senses. Over the ensuing decades, a great deal of scientific research has put teeth into their intuition, demonstrating just how important organic matter and micro-critters are in creating soil stability and nutrient cycling. The discovery of glomalin put a key piece of the puzzle in place. Now we know what we need: cover crops, undisturbed soil, deep plant roots, healthy populations of beneficial fungi, and glomalin. Lots and lots of glomalin, our unsung hero. It's time to sing its praises.

Which brings me back to abundance.

Here's a quote I like by Paul Feyerabend, who was a philosopher of science:

> The world we inhabit is abundant beyond our wildest imagination. There are trees, dreams, sunrises; there are thunderstorms, shadows, rivers; there are wars, flea bites, love affairs; there are the lives of people, Gods, entire galaxies.[2]

I like this quote because it reminds us that abundance exists at all levels and pretty much anyplace, if we choose to look. It challenges us to move beyond not just scarcity thinking, but also the surfeit of cheap, mass-produced, throwaway things generated by our consumer economy that we have come to equate with abundance. This type of abundance is an illusion, of course, built on a narrow, and narrowing, foundation of scarce resources. As Dorn's work demonstrates, we can get past this illusion. Look to nature instead, Feyerabend reminds us, for genuine bounty.

All of which raised a question in my mind: What would the world be like if we replaced scarcity thinking with the goal of creating as much real abundance as possible?

Dorn provided an answer after I had returned home from New England. He sent me an essay he wrote that included simple calculations regarding soil carbon. Since 1750, humans have transformed approximately 250 billion tons of solid carbon—in the form of organic matter, coal, and oil—into gaseous carbon dioxide, much of which is now polluting the atmosphere. What if we had turned those tons into new soil instead? Dorn calculated that at 10 percent organic matter, the soil would have been 6 inches deep across 2.3 billion acres of arid lands in northern Africa; at 46 percent organic matter, it could have been 8 feet deep across the 32 million acres of active cropland in the United States.

Can you imagine 8 feet of humus across America's girth? Can you imagine how much organic food we could grow with so much rich soil, or how much carbon dioxide we could drain from the atmosphere? Can you imagine that much abundance? I can't, but Dorn and his friends can.

And that gives me a lot of hope.

In his essay, Dorn writes that abundance thinking has "transformed the way I see myself as agrarian, and the way I seek to influence the balance of life in our soils, and my own balance of life." He now looks at the environment not as a source of scarce resources to be controlled and protected, but as a source of abundance with all the building blocks of life in place, including a bounty of carbon and nitrogen available in the atmosphere. By artfully exercising the same biological systems that create a productive patch of soil from a sandlot, he writes, we can change the way we think about our most basic of natural assets. Thinking about land this way changes everything.

He concludes:

> Agriculture should not be limited by statistics of current arable (plowable) land but a question of where land can be created and improved from abandoned lots and lawns, to eroded and depleted expanses that stretch across the horizon. The limiting factor of knowledge rather than of resources means that more people engaged within these systems is a path to healthy landscapes, people and society. A path which has yet to be built, but within our capacity. If we value carbon more than gold, we can all become alchemists.[3]

The Greenhorns are the new alchemists. Carbon and glomalin are the new alchemy. Food is the new gold. It's not an impossibility—it can be achieved, as Dorn and his colleagues are demonstrating. By combining high technology, systems theory, loads of data, organic no-till farming, social networking, barn dances, smiles, love, and compassion, these new alchemists aim to fulfill their vision of an abundant world, full of grains, sunflowers, laptops, and democratic values. It is exciting to see and share. There's a role for all of us too in this vision. We can sing the song of glomalin together where we live and work—everyone, everywhere. A song of prosperity, health, and happiness as a nation.

A song to ward off roving bands of giants.

Essential Minerals

COVER CROP WORKSHOP, EMPORIA, KANSAS

It must have looked silly.

Twelve of us were hunched over in a corn field under a blazing July sun, a few miles north of Emporia, Kansas, swishing butterfly nets among the corn stalks like deranged collectors chasing a rare breed of insect—deranged because it was a record-breaking 105 degrees! The federal government announced two days before I arrived that the Midwest was in the grip of the worst drought since 1956. Legions of farmers had begun plowing under or chopping up their stunted corn and soybean crops, already writing off the year as a complete failure. There we were, however, swishing our nets back and forth fifty times in a good-looking corn field owned and farmed by Gail Fuller, with nothing between us and the blazing sun except our determination to follow instructions and find spiders.

We found lots of spiders.

Back under the shade of a large oak tree, we handed our nets to our instructor, an affable entomologist with the US Department of Agriculture, who searched through them enthusiastically, pulling out spider after spider with his bare fingers (most spiders are poisonous, he told us, but very few can pierce human skin). Peering over his shoulder, I was amazed not only by the quantity of spiders in my net but by their diversity. I never knew so many odd-looking spiders existed! And who would have expected it from a corn field, in a record drought, during midday heat . . . which was exactly the point of the exercise, of course.

In a conventionally managed, monocropped Midwestern corn field, planted with genetically modified (GM) seeds, fertilized with industrially produced nitrogen, and sprayed with synthetic chemicals, there would be no spiders, the entomologist told us—drought or no drought. There wouldn't be much of anything living, in fact, except the destructive pests that could withstand the chemicals. The corn field we had just swept, however, was different, and I knew why. Fuller's field was no-tilled, it had a cover crop (and moisture in the soil as a result), it didn't use GM seeds, its corn coexisted with a diversity of other plants, and livestock were used to clean up after the harvest—all the things I had learned in my travels so far. All in one field, all under a broiling sun.

Seeing them together, however, wasn't the reason I had driven across humid Kansas in mid-July. I came to hear Jill Clapperton, an independent soil scientist and cover crop specialist, and to ask her a question: What happened to the nutrition in our food? And a second one: How can we get it back?

These questions first formed in my mind two years earlier, when I heard pioneering Australian soil scientist Christine Jones say at a conference that it was possible to buy an orange today that contained *zero* vitamin C. As in zilch. It got worse. In Australia, she continued, the vitamin A content of carrots had dropped 99 percent between 1948 and 1991, according to a government analysis, and apples had lost 80 percent of their vitamin C. She went on to say that according to research in England, the mineral content of nearly all vegetables in the United Kingdom had dropped significantly between 1940 and 1990. Copper had been reduced by 76 percent, calcium by 46 percent, iron by 27 percent, magnesium by 24 percent, and potassium by 16 percent. Furthermore, the mineral content of UK meat had dropped significantly over the same period as well—iron by 54 percent, copper by 24 percent, calcium by 41 percent, and so on.

This is important because all living creatures, humans included, need these vitamins and minerals to stay strong and healthy. Iron, for example, is required for a host of processes vital to human health, including the production of red blood cells (hemoglobin), the transportation of oxygen through our bodies, the conversion of blood sugar to energy, and the efficient functioning of our muscles. Copper is essential for the maintenance of our organs, for a healthy immune system, and to neutralize damaging "free radicals" in our blood. Calcium, of course, is essential for bone health. And every single cell in our body requires magnesium to function properly. Vitamins are organic compounds, by the way, composed of various chemicals and minerals, including carbon.

A deficiency or imbalance of these minerals (necessary to us only in small amounts) can cause serious damage to our health, as most people understand. That's why taking vitamin pills has become such a big deal—and big business—today, especially where young children are concerned. But few people stop to think about *why* we need vitamin pills in the first place. It's not simply because we don't eat our veggies, or because we drink too much soda, but because *the veggies themselves* don't have the amount of essential nutrients that they once did. As Jones quipped, for Aussies today to gain a comparable amount of vitamin A from carrots that their grandparents could, they'd have to eat themselves sick.

How did this happen? Well, the quick answer is that industrial agriculture happened. The hybridization of crops over the decades for production values—yield, appearance, taste, and ease of transport—has drained fruits and vegetables of nutrients. But the main culprit is what we've done to the soil. As a consequence of repeated plowing, fertilizing, and spraying, the top few feet of farmland soil has been (1) leached of its original minerals and (2) stripped of the biological life that facilitates nutrient uptake in plants. Some farms, especially organic ones, resupply their soils with mineral additives, but many farms do not, preferring to rely on the Big Three—nitrogen, potassium, and phosphorus (NPK)—to keep the plants growing. According to the industrial mind-set, as long as crops are harvestable, presentable, digestible, and profitable, it doesn't matter if their nutrition is up to par. If there's a deficiency, well, that's what the vitamin pills are for!

However, it was the next thing that Jones said that spun my wheels. There was another way to remineralize our bodies without having to rely on pills or their corporate manufacturers: restore essential elements the old-fashioned way—with plant roots. With carbon, specifically. Building humus by increasing the amount of carbon in the soil via no-till agriculture, planned rotational grazing, and other practices that stimulate mycorrhizal fungi/root activity and the production of glomalin, she said, would (1) increase the availability of potassium, calcium, phosphorus, sulfur, copper, zinc, iron, magnesium, and boron to plant roots (which are good for plants); (2) reduce availability of sodium and aluminum (which are bad for plants); and (3) increase the pH in the soil (from acidic to neutral—good for everything).

Access to these essential minerals in combination with carbon means vitamins and other types of nutrients, including acids, carbohydrates, fats, and proteins, can be produced within a plant.

One key to building soil carbon on farms is *cover crops*—plants that keep the land covered with something green and growing at all times, even in winter. As I had learned from Dorn Cox, cover crops = deeper roots = greater carbon production and better carbon storage. I also learned from Tom and Mimi Sidwell and other livestock producers that native grass was a form of perennial cover crop. What wasn't quite clear to me, however, was how cover crops put essential minerals/nutrients back in our food. So I went to Kansas to find out.

Clapperton, who hails originally from Canada but lives today on a Montana ranch, told the workshop audience that the key to rebuilding soil health is to start a "conversation among plants." Cool-season grasses (such as barley, wheat, and oats) and cool-season broadleaf plants (such as canola, pea, turnip, lentil, radish, and mustard), she said, need to dialogue constructively with warm-season grasses (including millet, corn, and sorghum) and warm broadleafs (such as buckwheat, sunflower, and sugar beet). Who gets along with whom? Who grows when? Who helps whom? If you can get these plants engaged in a robust conversation in one field, she said, you'll be creating "a feast for the soil." That's because increased plant diversity, as well as year-round biological activity, absorbs more CO_2, which in turn increases the amount of carbon available to roots, which feeds the microbes, which builds soil, round and round.

This is exactly what happened on Fuller's farm. When he took over the operation from his father they were growing just three cash crops: corn, wheat, and soybeans. Today, Fuller plants as many as fifty-three different kinds of plants on the farm, mostly as cover crops, creating what Clapperton called a "cocktail" of legumes, grasses, and broadleaf plants. He doesn't apply any herbicides, pesticides, or fertilizers either, despite the recommendations of his no-till neighbors and chemical manufacturers who advise them. That's because Fuller considers "weeds" to be a part of the dynamic conversation as well. Besides, chemicals kill life, Clapperton reminded us, including spiders, dung beetles, and even grasshoppers.

As a result of this big, robust conversation, Clapperton said, the carbon content of the soil on the Fuller farm has doubled from 2 percent in 1993 (when they switched to no-till) to 4 percent today. That's *huge*. But what about the mineral content of Fuller's crops?

That's risen dramatically too, she said, and it's done so for two reasons: First, no-herbicide/no-pesticide no-till means the microbial universe in the soil remains intact and alive, and if the soil dwellers have enough carbon (as an energy source) they will facilitate the cycling of minerals in the soil, especially earthworms, who are nature's great composters. Second, a vigorous and diverse cover of crops will put down deeper roots, enabling plants to access fresh minerals, which then become available to everything up the food chain, including us. And by covering the soil surface with green plants, or litter from the dead parts, Clapperton said, a farmer like Fuller traps moisture underground, where it becomes available for plants and animals (of the micro variety), enabling roots to tap resources and growing abundant life.

"Aboveground diversity is reflected in belowground diversity," she said. "However, soil organisms are competitive with plants for carbon, so there must be enough for everybody." Predator-prey relationships are also important to nutrient cycling, she said. Without hungry predators, such as protozoa and nematodes, the bacteria and fungi would consume all the nutrients in the soil and plants would starve. Predators aboveground play a positive role too, including spiders and especially the number one predator, ants!

So exactly how do minerals get into plants? There are two principal paths: First, minerals can dissolve in water, and when the water is pulled into the plant through its roots, the minerals are absorbed into the cells of plant tissue. Whichever minerals the plant doesn't need (or doesn't want) will remain stored in the cells. Second, mineral nutrients can enter a plant directly by being absorbed through the cell walls of root hairs. Some minerals, such as phosphorus, can also "hitch a ride" with mycorrhizal fungi, which then "barter" them for carbon molecules from the plant roots. Of course, if there aren't any minerals in the vicinity, no uptake into plants is possible!

It all begins with a dynamic conversation at a cocktail party for plants—where everyone is gossiping about carbon!

Standing under the oak tree at the end of the workshop, after we had oohed and aahed over a giant wolf spider someone discovered under a shrub, Clapperton reminded us why using nature as a role model—for cover crops in this case—was so important: we need to recycle nutrients, encourage natural predators to manage pests, and increase plant densities to block weeds, which in a natural system are all integrated and interconnected strategies.

This reminded me of something the great conservationist Aldo Leopold once wrote:

> The black prairie was built by the prairie plants, a hundred distinctive species of grasses, herbs, and shrubs; by the prairie fungi, insects, and bacteria; by the prairie mammals and birds, all interlocked in one humming community of cooperations and competitions, one biota. This biota, through ten thousand years of living and dying, burning and growing, preying and fleeing, freezing and thawing, built that dark and bloody ground we call prairie.[1]

One biota. With carbon at its core.

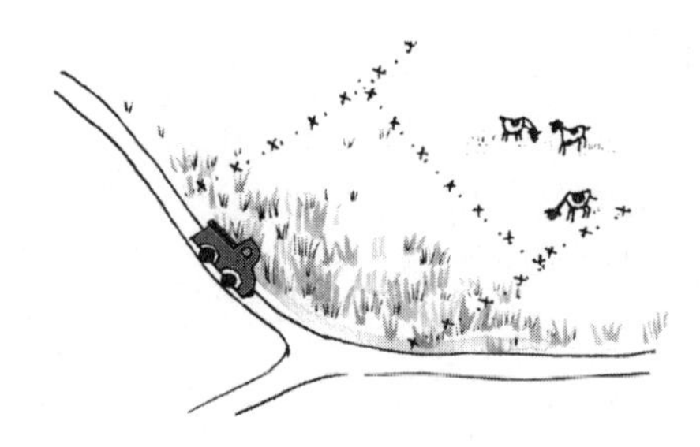

3

COEXISTENCE

Colin Seis's first attempt at breakfast didn't go well, a product of too much heat under the eggs and too much distracting conversation about saving the world.

Colin was responsible for the eggs, but the distracting conversation was my fault, so I kept quiet on the second go and in short order a delicious breakfast was ready. I was a guest on Colin's farm, located 200 miles northwest of Sydney, Australia, in order to learn about a revolutionary agricultural practice called *pasture cropping*. Since the late 1990s Colin has been no-till drilling an annual cereal crop into perennial pasture on his sheep farm, which means Colin produces two crops from one parcel of land: a cereal crop for food or forage *and* wool or lamb meat from his pastures. Not only does this represent a complete break from agricultural tradition, which firmly believes that crops and pastures should be kept separate from one another, it raises the possibility of feeding large numbers of people in a manner harmonious with nature's way of doing things. With the world's population projected to rise from seven billion to nine billion by 2050 (and to keep rising), this possibility is significant—thus the distracting conversation and the burned eggs.

The link is carbon—and I'll come to that in a minute.

In 1993, Darryl Cluff, a farmer and neighbor of Colin's, asked himself a question: why are crops and pastures farmed separately? The answer: tradition. He had been taught that pasture and crop systems operated by different ecological processes and were thus incompatible. Furthermore, tilling eliminated the grass that animals required. The systems could be alternated over the years, but never integrated. Coexistence was impossible. Right? Or wrong? Cluff asked the question because he had been observing the interplay between annual and perennial plants on his farm and had begun to wonder why they were kept apart by humans when nature apparently had no such hesitation. Nature certainly wanted *weeds* on his farm, he noted, so why not encourage a more useful type of annual instead, such as oats? Cluff had a guess: weeds liked to run a 100-yard dash while perennial grasses like to a run a marathon. Two different races, two different types of athletes.

Right? Or wrong?

What if it were just one race? What if pasture acted as a kind of cover crop for the annuals, keeping down the weeds but allowing the middle-distance runners, such as oats or barley or canola, to grow while the perennials waited for their turn on the racetrack? More to the point: what if you no-till drilled the perennial pasture during its dormant period with a cereal crop? Why couldn't a cereal plant be cropped in a perennial pasture? And once the crop was harvested, couldn't the sheep go in and graze afterward? Couldn't farmers (they don't call them "ranchers" Down Under—everything is a farm unless it's huge, and then it's called a "station") figure out a way to make annuals and perennials get along symbiotically? Couldn't they grow a crop in a pasture? If nature could do it, why couldn't they?

Cluff decided to give it go. In 1995, he drilled his first crop of oats in dormant native perennial pasture and, despite dry conditions that year, he was able to harvest a decent crop. Pasture cropping was born! The following year he tried wheat, again with good results. In 1997, he and Colin wrote a paper about his experience for a national conference with the odd but wonderful title "Should Farmers and Graziers Be Garmers and Fraziers?" Possibly sensing the unmarketability of these terms, Cluff coined the term *pasture cropping* instead to describe what he had been accomplishing on his farm. In 2003, he gathered his knowledge and experience about pasture cropping into a book with the quirky title *Farming without Farming*—but there was nothing quirky about its contents.

In the meantime, both Colin and Christine Jones had been following Cluff's work closely and were inspired by his results. In 1996, Colin began pasture cropping on his farm for the first time, with good success. As neighbors, he and Cluff often discussed their progress the Aussie way—over beers. For her part, as a soil scientist Christine understood the ecological processes at work on Cluff's and Colin's farms, including pasture cropping's beneficial effect on the carbon cycle. She reported her findings in various research journals and conferences, including a big agricultural event in Washington, D.C., in 1998. Cluff's idea seemed poised to take off, despite stiff resistance from the traditionalists. But new ideas can have a long gestation. Colin began waving pasture cropping's banner, organizing workshops and speaking at conferences, and when I heard about his work, I knew I had to visit his farm, pronto.

But let's back up for a moment and put things in a historical perspective.

When Colin talks about pasture cropping he likes to start with the story of what went wrong with Australian agriculture. It's his way of placing pasture cropping in context, as well as explaining why he calls what he does *regenerative agriculture*—because so much of what happened on the continent's farms historically has proven to be decidedly unregenerative.

The story begins in the 1780s when old Mother England, after losing her colonies in the American Revolution, began looking for a new dumping ground for her overflowing prisons. Explorer James Cook's positive report of a fertile and temperate land at the other end of the world, soon to be called Terra Australis, which he claimed for the king in 1770, provided just the opportunity the empire was seeking.

By 1791, two fleets of thieves, poachers, counterfeiters, highwaymen, and troublesome lasses—along with a forlorn contingent of soldiers, officers, and one highly competent governor (luckily for them)—had carved a home from the wilderness in Sydney Cove. Soon, they were starving. With enterprising grit, however, the colony survived and quickly overspread the continent in an energetic rush of discovery and settlement not unlike the tide of manifest destiny that carried settlers across the vast American frontier. What began as a penal outpost had by midcentury become one of the last great jewels in the English crown. But with typical independent spirit, by 1902 Australia was a sovereign nation, well on its way to becoming the prosperous, gregarious, beer-loving country that we know today.

Not so familiar is the story of the land itself. The continent is salty, flat, and mostly dry. Repeated submerging by the ocean over the eons combined with a lack of mountain-making geological uplift (necessary for the weathering of rock into life-giving topsoil) created thin, nutrient-poor soils that were rapidly depleted by a pattern of colonial settlement and agricultural use designed for the rich, wet climes of England.

The destruction of Australia's grasslands began with inappropriate grazing management, said Colin, and expanded later with extensive plowing. Overgrazing, tilling, and the introduction of nonnative animals (including foxes, rabbits, and toads) and a variety of aggressive plant species all combined to devastate the continent's soils and nearly defenseless indigenous wildlife—as well as its indigenous people—in less than a century. Topsoil began to wash away, along with its precious carbon content, causing a general decline in overall soil health and crop productivity. Everything sped up with the introduction of the mechanized tractor in the 1920s, and not in a good way. This was followed by widespread application of herbicides, pesticides, and chemical fertilizer in a desperate attempt to salvage what remained of the soil's fertility.

Colin knows this story firsthand because he saw it happen on his family's 2,000-acre farm in New South Wales, called Winona.

Colin's grandfather resisted the industrial changes being pushed on Australian wheat farmers by agricultural companies and government agencies. However, his son, Harry, decided to try something called "super manure"—an early version of superphosphate—in an attempt to boost declining yields. Harry's father objected,

Colin said, asking "What's wrong with the old manure?" Trouble escalated when Harry bought a tractor. He didn't know it, but his increased plowing depleted the soil, of carbon especially. A vicious cycle ensued: tilling meant less fertility in the soil, which required more chemical inputs, which was followed by more tilling, round and round. Then the farm began to fail. Costs kept rising, fertility kept falling, salinity spread, trees began to die—and Colin's family began to go broke.

"Still, the 'moron' principle prevailed in my family," said Colin as we finished breakfast, his voice rising slightly, "you know, more on and more on."

The farm ended up becoming dysfunctional and unprofitable. The farm's granite soil had become compacted and acidic and organic carbon levels had dropped to below 1.5 percent. The topsoil had been reduced to less than 100 millimeters (4 inches) in thickness, and pockets of salinity were breaking out around the property.

In 1979 came the kicker—a wildfire burned almost all of Winona. Three thousand sheep died, all of the buildings were destroyed, 20 miles of fencing burned up, trees exploded, grass died, and Colin ended up in the hospital being treated for burns.

"Worst of all, there was *no* money to recover things with, which means we had hit rock bottom," Colin explained. "My grandfather had the last laugh, I'm afraid."

When Colin had recovered from his burns, he decided to rethink the way he had been practicing agriculture. It wasn't a criticism of Colin's father (who had followed the rules of farming for the time, Colin said), but rather a realization that the rules themselves needed to change. The fire suddenly created an opportunity to do just that. Colin vowed that out of the ashes a new farm would emerge.

The first step was to physically rebuild the farm, which took two years and a lot of help from neighbors. The second step was to go cold turkey on fertilizer, herbicides, and pesticides. The pastures collapsed as a consequence, Colin said, because they were addicted to chemicals. This led to the third step: research the farm's native grasses. Could they come back? If so, could they be as productive as the nonnative species? Colin said his father had fought against native grasses all his life, yet they kept returning despite his efforts at eradication. This raised a question in Colin's mind: if they keep wanting to come back, why not let them? Apparently they wanted to be there.

This led to the fourth step: study the holistic management ideas of Allan Savory, who had developed a way of managing livestock on pasture that mimics the graze-and-go behavior of wild herbivores. Colin resisted this step initially because Savory's ideas were seen as controversial among his farming colleagues, but in the end he felt that he had no choice. He quickly learned that the system worked, especially when he turned his sheep loose on the nonnative plants (with his father's reluctant blessing). However, this new approach to grazing management created a long transitional period of low productivity, which reinforced his neighbors' belief that native grasses were not as productive as introduced ones. Colin persisted with his plan, however.

"I'm stubborn like my dad and his dad," Colin said. "I wasn't sure if that was a good thing or not for a while, but in the end it paid off."

By 1990, things had improved substantially, and Colin was seeing benefits both on the land and in his bank account. But he knew it wasn't enough to completely repair all the damage that Winona had endured over the decades. He needed a new idea, something revolutionary.

"Before industrialized agriculture was developed, the world's grasslands and farms contained hundreds of plant species of all sorts," Colin said. "And they functioned well together with very few problems like disease, insect attack, and weeds because it was a balanced ecosystem. Pasture cropping returns that balance. It also creates good, rich soil with high carbon levels and good water-holding capacity."

It's worked, and Colin wanted to show me the result. So we cleaned up the remains of breakfast, climbed into his truck, and took a drive across the farm. I immediately leaned out the window of the vehicle in order to get a good view of the kangaroos that seemed to be everywhere. They looked both exotic and magnificent to me, but when I asked Colin about them, he just shrugged. Many farmers, he said, consider kangaroos to be pests, partly because there are so many of them and partly because they're grazers, which means they compete for forage resources with livestock. "I don't mind them, though," he said. "They're beautiful creatures, and thanks to our grazing management we've figured out a way to get along alright on the farm."

I leaned out the window once again and was struck this time by the beautiful landscape. The gently rolling hills were covered in grass and festooned with clumps of lovely eucalyptus trees (in their native habitat) as far as the eye could see. It was a bucolic scene in every sense of the word. It also looked like ideal sheep country—which it is, thanks in large part to a condition unfamiliar to Americans: no predators. No wolves, coyotes, bears, cougars, lynx, or bobcats. The only wild, sheep-killing native predator in Australia is the dingo dog. However, none lived near Winona, Colin said. Foxes are a problem occasionally, he continued, but in general his herd of Merino sheep graze in peace.

We drove past his son Nicholas's house to the edge of a large field. Parking, we climbed out and strode purposefully into the pasture, aiming toward what I could already see were oat plants growing in neat rows among the grass. Although it was early in the season and rains had been stingy, the plants looked vigorous. Stopping at the first row, Colin explained how pasture cropping works.

More important than the relationship between annual and perennial plants in a field, he said, is the relationship between C3 (cool-season) plants and C4 (warm-season) plants—the "C" in this case standing for carbon. The difference between the two types is the number of carbon molecules each possesses and how they influence the process by which glucose is produced in a plant. C3 plants, such as wheat, rice, oats, barley, and some grasses, grow early in the season and then become less active or go dormant as temperatures rise and sunlight intensity increases. In contrast, C4 plants, such as corn, sorghum, millet, and many pasture grasses, remain dormant until temperatures become warm enough to "switch them on" and they begin growing.

Pasture cropping utilizes the niche created by C3 and C4 plants. When a C4 grass plant is dormant (during winter), a C3 crop seed is sown by no-till drilling into the C4 pasture. With the onset of spring, the C3 plants begin to grow. If managed properly, and given the right amount of rain, the C3 crop can be harvested before the C4 plants begin the vigorous part of their growth cycle. The removal of the C3 crop will then stimulate C4 plant growth (due to reduced competition). Also, the shallow- and deep-rooted plants access water and mineral resources in the soil differently, which can reduce competition and increase overall productivity.

A key is what's happening in the soil. C3 cereal crops provide sugars to soil microbes, such as fungi, nematodes, and protozoa, during the time when the C4 plants are dormant. This builds up soil carbon and improves fertility faster than a C4 pasture alone could. It also speeds up nutrient cycling, promotes an improved water cycle, increases nitrogen content, and adds organic matter to the soil—which builds humus. Additionally, the no-till drill lightly aerates the soil, allowing oxygen and water to infiltrate to deeper levels than they might otherwise.

Another key is using grazing animals to prepare the C4 field before drilling. If grazing animals hit the perennial pasture hard in the late fall, they give the C4 plants a "headache," as Colin puts it, so that the C4 plants come up more slowly in the spring, giving the C3 plants a chance to grow. By hitting the pasture with a large mob of sheep in a time-controlled manner, Colin can keep the C4 plants from growing too tall too early and thus prevent them from shading the C3 plants. Animals can also control weeds, create litter on the soil surface, supply a pulse of organic nutrients (manure and urine) for the crops, and remove dry plant residue from the pasture.

Colin said that his use of sheep "mobs" has been controversial in some quarters due to a concern about soil compaction. This is only a problem, though, where there are low levels of ground cover and litter, he said, or when the ground is very wet. "Where there are good perennial pastures and ground cover, pasture-cropped paddocks show very little compaction and soil structure problems."

Proper sowing technique is another key, as is an assessment of a pasture's potential before a farmer tries to crop it. Colin has some advice: before sowing, graze the paddock down, create as much litter as possible, use an herbicide to control weeds only when absolutely necessary, use no-till equipment to sow at the correct depth and row spacing, sow the correct crop for your soil type, conduct a soil test if possible, and avoid fertilizer use as much as possible—it shouldn't be necessary!

One more instruction: "Never, never, never use a plow."

Or light a fire. Don't burn anything. "Throw your matches away," he said. Use livestock instead to knock down the plants.

Needing to stretch our legs a bit, we decided to walk across the field. As we strolled, Colin listed the various reasons he continues to give pasture cropping "a go," as he put it:

- It's profitable—Colin and his son run around 4,000 Merino sheep and pasture-crop around 200 hectares (500 acres) annually in oats, wheat, and cereal rye.
- The farm has steadily improved its sheep carrying capacity, wool quality, and wool quantity.
- Winona is now almost entirely native grassland, with over fifty different species of grasses, forbs, and herbs.
- The farm saves around $60,000 annually in decreased inputs (fertilizer, etc.) in comparison to its former operation.
- Crop yields from pasture cropping remain about the same when compared to conventional cropping, with oat yields averaging 2.5 tons/hectare.
- Insect attacks and fungal diseases in crops or pasture are minimal.
- There has been a noticeable increase in bird and native animal numbers, as well as species diversity.
- Soil microbial counts show that the Winona soil has significantly higher counts of fungi and bacteria than before.
- According to a soil analysis, all minerals and nutrients have increased by an average of 150 percent.
- Perhaps most impressively, soil carbon has increased by 203 percent over a ten-year span compared to an adjacent farm (owned by Colin's brother). Christine Jones calculated that 164 tons of CO_2 per hectare has been sequestered to a depth of 0.5 meter on Winona. This has contributed to a dramatic increase in the water-holding

> capacity of the soil, which, according to Jones, has also increased by 200 percent in ten years and can now store over 140,000 liters/hectare of extra moisture.

Wow. It's no wonder that pasture cropping is practiced today by over two thousand farms across Australia, according to Colin.

But there was more.

Colin calls what he and his son do economically on Winona "vertical stacking"—that is, the stacking of business enterprises so that they fit together in a way to generate more profit per hectare regeneratively. Pasture cropping is a perfect example. It lowers the cost of growing crops to a fraction of conventional cropping methods; it adds six months' extra grazing forage compared with the loss of grazing due to ground preparation and weed control required in traditional cropping methods; and it recruits perennial plant diversity naturally, which means that there is no need to resow pastures—a cost that can run from $100 to $150 per hectare.

"The best way to improve your profits is to improve your soil," Colin said.

Pasture cropping is also an effective land restoration strategy, which is precisely how Colin used it on Winona—to convert a worn-out, weed-dominated, burned-over, failing patch of farmland into an ecologically healthy and economically profitable landscape. Colin is convinced the same result can happen anywhere similar C3–C4 plant relationships exist (in other words, where pastures have a true dormant season).

"It's a great way to rebuild grasslands and can happen almost anywhere there's enough rain to grow a crop," he said.

What about pasture cropping in arid environments? When I asked, Colin replied that you must no-till drill more carefully and expect yields to be lower, especially in the first few years. What about going organic? It's certainly possible, he said, but he hasn't tried yet, preferring to "keep every tool in the toolbox." What about reducing fossil fuel use? (Colin uses a big harvester to bring in the crop.) "It can also be done with horses or electric engines, I suppose," he said. "You're only limited by your imagination."

Back at the house, Colin gave me a copy of a 2010 research study conducted by the University of Sydney, under the direction of Peter Ampt, that compared Winona to the adjoining farm owned by Colin's brother. (Colin says his brother has been "a good sport" about this.) The goal of the research was to evaluate the effects of pasture cropping versus conventional management on soil health and ecosystem function. The project compared paddocks of comparable size on each farm. Here are some of the results of the research:

- Winona's paddock was 83 percent native perennial grass species.
- His brother's paddock was 41 percent annual grass species.
- Soil microbial counts showed that Winona had higher amounts of fungi and bacteria than the neighboring farm.
- Winona had higher levels of soil organic carbon.

In the study's conclusion, Ampt and his coauthor Sarah Doornbos write:

> These results illustrate that the rotational grazing and pasture cropping practiced on the innovator site can increase perennial vegetative ground cover and litter inputs, compared to the continuous grazing system and conventional cropping practiced on the comparison site. Increased perenniality and ground cover lead to improved landscape function in the pasture through increased stability, water infiltration and nutrient cycling which in turn can lead to improved soil physical and chemical properties, more growth of plants and micro-organisms and an ultimately more sustainable landscape. It also shows that rotational grazing and pasture cropping can improve landscape function while sustaining similar or higher stocking rates over the year compared to the conventional system.[1]

There was one more benefit to pasture cropping—the one that had caused Colin to burn our breakfast eggs that morning: it has the potential to feed a lot of people.

Knowledge that there will be nine billion people on the planet by 2050 raises a profound question: how are we going to feed them without destroying what's left of the natural world, especially under the stress of climate change? If humans can't find enough food, fuel, fiber, and fresh water to ensure their well-being, they'll raid the environment to secure these necessities, pushing many other values that we place on nature down the priority list.

It's not about poor people either. The food well-fed Americans and Australians eat comes from a global production system that is already struggling to find enough arable land, adequate supplies of water, drought-tolerant plants, and hardy animals to feed seven billion people today. Add two billion more and you have a recipe for a devastating raid on the natural world. Where is all this extra food and water going to come from, especially if the climate gets hotter and drier in many places as predicted?

Industry has an answer: the moron principle. More chemicals, more fertilizers, more GM organisms, more monocropping, more heavy fossil fuel use. Then there is industry's idea of "innovation." I read recently that industry chemists are attempting to "reengineer" photosynthesis in order to make plants "more efficient" (and patentable by corporations, undoubtedly). A second "green revolution" is required, industry leaders say, even though the consequences of the first revolution have been decidedly mixed.

Fortunately, there is another way—a regenerative way—as pasture cropping demonstrates. Not only does it restore land and people, it can *intensify* food production sustainably. On Winona, Colin produces a grain crop and an animal product—*plus* (if he wanted to) a wild harvest crop of grass seeds, traditionally used as a food source by the Aboriginal people of the area—all from the same hectare, while at the same time building topsoil, improving the water cycle, enhancing the nutrient quality of the plants, and restoring land health. All carefully integrated and managed under Colin's stewardship.

Talk about win-win-win. All we needed now was a beer to celebrate!

It was the power of carbon + coexistence at work, a realization reinforced a few days later when I stopped to visit another farm in New South Wales, where, I had been told, the number of native grass species had increased from 7 to 130 in only seven years! The key? Using cattle and sheep managed together as one herd.

The man responsible for this accomplishment was Eric Harvey, a gregarious former wool trader who had decided to try his hand at the other end of the supply chain by purchasing a 7,000-acre farm called Gilgai, located a few miles from the crossroads city of Dubbo. Shortly after buying Gilgai in 2004, however, Eric nearly "bought the farm" himself when he had a massive heart attack, as he explained to me on the drive in from Dubbo. After recovering, Eric was astonished to learn from his doctor that his body was almost completely devoid of minerals, which are essential to human health. He knew there weren't many minerals in rainwater—due to water scarcity Australians collect and drink a lot of rainwater—but he assumed he was getting enough minerals from the plants and animals he ate, which in turn get their minerals from the soil. Ninety-five tests showed he wasn't. This was a huge eye-opener, he said.

Eric had soil tests conducted at Gilgai, discovering that it too was depleted of essential minerals, including carbon. This meant that the farm and Eric's health were now one and the same—both had to recover. But, he wondered, where were the minerals going to come from? A mine? A factory? That didn't sound very practical or economical. And what about carbon? Was he supposed to spread compost over all 7,000 acres of land? That didn't sound economical either.

A chance conversation with a neighbor provided Eric with an unexpected answer: the sky. Carbon was freely available in the air, his neighbor said, in the form of carbon dioxide, and all Eric had to do was get it into the soil via photosynthesis, livestock, and planned grazing practices. The goal, he told Eric, was to grow native grass—diverse and copious amounts of it.

So that's what he did. First he studied the principles of planned grazing, and then, after deciding to put them to work, he made another unconventional decision: to run cattle and sheep together as one grazing unit. It's called a *flerd*—a flock of sheep and a herd of cattle, commingled. Years ago he had seen sheep and cattle grazing on a farm in Africa and thought, "That makes sense." Maybe to Eric—but not to many others. To say that it is not traditional to run cows and sheep together would be a huge understatement. It's hardly done anywhere. Not only do many in agriculture consider the two types of herbivores to be incompatible with each other from a grazing perspective, but most sheep and cattle farmers consider *each other* to be incompatible as well. In fact, Australia endured its share of range wars between sheepmen and stockmen over the decades, much like America did in the nineteenth century.

Eric ignored all that, and in 2005 he put together his first flerd, eventually commingling five thousand sheep and six hundred cows. His goal was to use the different grazing behaviors of sheep and cattle to benefit plant vigor, diversity, and density. Nature likes mixed-species grazing, Eric said, because animals often complement each other in what they will eat, the composition of their manure, and the way their hooves interact with the soil. As Eric described it, herbivory creates an organic "pulse" *below* the ground surface as roots expand and contract with grazing. This feeds carbon to hungry fungi, protozoa, and nematodes, which in turn feed grass plants. The manure "pulse" aboveground helps too, especially with nutrient cycling. His plan with the flerd was to make both "pulses" beat stronger and more steadily.

To accomplish this goal, Eric divided the 7,000-acre farm into 196 paddocks, mostly with electric fencing, creating an average paddock size of 140 acres (the smallest is 6 acres). The flerd moves from paddock to paddock every few days, giving each paddock plenty of time to grow more grass. And with only one "mob" to watch, Eric is often back home by 10 a.m. As further work reduction, Eric monitors the watering troughs remotely via sensors linked to the computer in his office, as he showed me, which supply up-to-the-minute data. He also pays for a service that provides aerial infrared images of his farm daily, which allows Eric to monitor the growth rate in his paddocks at a 7-acre scale. He calls this service "pastures from space" and says it gives him an invaluable snapshot of forage conditions, which helps him adjust his grazing schedule.

Eric also ground-truths the monitoring data he receives. That's how he knows he has been able to expand the number of plant species on Gilgai from 7 to 130. This improvement in diversity has substantially enhanced the mineral content of the plants, since they can now access nutrients more widely, as well as deeper in the soil profile, and process them more effectively. And when these plants are eaten by animals, which are in turn eaten by us, the minerals enter our bodies, as Eric can personally attest (his physical health has improved dramatically). That's why Eric and his family grow and sell only grass-fed products from their farm. By definition, grass-fed means an animal has spent its entire life on grass or other green plants, from birth to death. This contrasts with the feedlot model in which an animal finishes its life in confinement, fattened on grain and assorted agricultural by-products and pumped full of medication and other chemicals.

Thanks to a lot of digging in the scientific literature over two decades by Jo Robinson, an independent researcher in America, the health benefits of grass-fed over feedlot meat have become widely known. They include the following:

- more omega-3 fatty acids ("good" fats) and fewer omega-6 ("bad" fats)
- fewer saturated fats linked with heart disease
- much more conjugated linoleic acid (CLA), a cancer fighter
- much more vitamin A
- much more vitamin E
- higher levels of beta-carotene
- higher levels of the B vitamins thiamin and riboflavin
- higher levels of calcium, magnesium, and potassium
- positive effect on enhancing immunity, increasing bone density, and suppressing cancer cells
- does not contain traces of added hormones, antibiotics, or other drugs

As Robinson says, "If it's in their feed, it's in our food"—which means it's in us.[2]

In the past few years, another important advantage of grass-fed meat production has emerged: it has a smaller carbon footprint. By some estimates, meat from grass-fed animals requires only 1 calorie of fossil fuel to produce 2 calories of food. In contrast, feedlot beef requires 5 to 10 calories of fossil fuel for every calorie of food produced. The big differences include the fertilizer and other inputs used to grow the corn feed and the amount of transportation involved in placing feedlot beef in supermarkets across the nation.

The carbon footprint advantage has been challenged by some experts, however, who claim that methane emissions are higher with grass-fed livestock, and that overall impacts on land health and water quality (due to overgrazing) are fewer with feedlots. Disagreeing with these experts, however, is a report by the Union of Concerned Scientists (UCS), which claims that the overall greenhouse gas impact of grass-fed livestock is *positive*. Well-maintained pastures and careful management of grazing animals can draw greenhouse gases out of the air and store them in the soil, where they fuel plant growth. Feedlots have no living plants, the UCS notes, just bare dirt and manure. Instead of absorbing greenhouse gases, as healthy grasslands do, they emit them.[3]

Back at Gilgai, as for the flerd itself, Eric has hardly had any trouble running sheep and cattle together. The key is to raise them as one family, he said, especially the lambs. Sheep will bond with cows at a young age and remain bonded for the rest of their lives. As a result, the sheep follow the cattle wherever they go, which means they'll move from paddock to paddock with the herd without much fuss. This is great news for a multipaddock farm like Gilgai. It also means Eric doesn't have to train any sheep to electric fencing, only the cattle. "Needless to say, moving one herd of livestock is a lot easier than moving two," he said. "You just make sure there's enough forage and water ahead of them."

The only trouble he's had, other than an occasional grumpy cow who doesn't like sheep—quickly culled—happens during calving, when mama cows become highly protective and might kill an ewe that comes too close. Eric solves this by separating the cattle from the sheep during their respective birthing seasons. "The only other conflict I've ever seen is over shade," says Eric. "And that's been minor. Otherwise, they get along great."

We went to see for ourselves. After a quick stop for a look inside a sheep-shearing shed (which I had only seen in Australian movies), Eric and I walked down a dirt lane, crossed through a gate, and entered a grassy field. The cattle saw us coming. A number of them jogged hopefully toward us until it became clear that we weren't going to open a gate so they could move to fresh grass. They drifted off, followed closely by small flocks of sheep. We stopped in the middle of the paddock. Looking around, I saw cattle and sheep *everywhere*. "Look how they spread themselves out," Eric said. "Cattle prefer grass over forbs [broadleaf plants], but it's vice versa with the sheep. If you keep them in a paddock just the right amount of time, everything gets a nibble. That's good for the plants and the soil."

"They'll all be out of here tomorrow," he added.

Although Eric doesn't run goats as part of the flerd, he said there's no reason it couldn't be done. Not only do goats get along with sheep and cattle just fine, but, if bonded properly, goats prefer brush and weeds over grass and forbs, which means they would add another level of grazing diversity to a pasture—while also being good for the soil.

According to some research I had done prior to my trip, another benefit to a flerd is protection from predators, such as coyotes. In the American West, coyotes are the scourge of sheep, lambs especially, which is one reason why sheep-only ranching has declined steadily over the decades as predator populations rebounded, wolves especially. Experiments, however, have shown that when sheep are bonded to cattle they are protected from predation by coyotes, which are reluctant to take their chances with a closely packed herd of bovines. Experiments have also demonstrated that sheep gain weight faster when grouped with cattle compared with sheep that are managed as a separate flock. Wool production was also greater with the flerd than with sheep foraging alone—a fact that Eric said he could confirm. He attributed both improvements to the healthier soil and increased diversity of plants on Gilgai—a result of his careful stewardship.

"How many other farmers in Australia have a flerd?" I asked Eric as we walked back to the shearing shed.

"We're the only one I know of," he replied.

"Wow. So the belief that cattle and sheep don't mix still persists despite all the signs to the contrary?" I said.

He nodded.

I mentioned to Eric a quote from an animal scientist that I had found during my research: "The biggest challenge to multi-species grazing is getting beef people to not laugh when you recommend they get some sheep."

"It's a paradigm thing with humans," Eric said. "It's not an issue in nature."

Lastly, as with Winona, coexistence on Gilgai is profitable. Beef, lamb, mutton, and wool are all produced on the farm—all from one flerd. Although we didn't get into a "saving-the-world" conversation as I did with Colin, I thought the type of sustainable food intensification that Eric has accomplished on his farm was as potentially consequential as it was on Winona. And just as hopeful.

After my return to the United States, Eric e-mailed me to say that a new survey of Gilgai had found six additional plant species on the farm, raising the total from 130 to 136.

He sounded like a proud father.

Coexistence was clearly a major reason for the success of both Winona and Gilgai, but the real link between the two farms, I realized, was carbon. Both farms were worn out from past management, which meant their carbon cycle had been severely damaged and the quantity of this essential substance badly depleted. But I now understood that carbon is nature's "comeback kid." After all, its source—the air—is free and profoundly abundant. Thanks to new ideas and a willingness to take chances (augmented by beer), Winona and Gilgai, as well as their respective owners, had their health restored and were productive once more. By managing the carbon cycle the old-fashioned way, with animals, plants, and microbes, Colin and Eric had not only restored themselves and their land, they were in a position to potentially feed the world regeneratively. Coexistence was crucial to their success, but so was carbon. Which raised a question in my mind: where did this "comeback kid" come from, anyway?

When the modern carbon-rich cornucopia we call Earth got its act together 4.5 billion years ago, it had relatively little of the life-giving element lying around, scientists say. Carbon only arrived in useful quantities over the millennia as the result of a steady bombardment by carbon-rich comets and asteroids. But where did they get *their* carbon? And what about the sun and Jupiter, both of which are also carbon-rich? Where did this life-giving element come from, and how did it get here?

To answer these questions, we must go back to the origin of the universe.

In the beginning, there was no carbon. When the Big Bang created the universe approximately fourteen billion years ago, the explosion was so phenomenally hot that all matter existed solely as protons and neutrons. Then, one minute after the Big Bang, when the temperature dropped to "only" one billion degrees Celsius, some neutrons began to decay into hydrogen, while others collided with protons, creating helium. These were the two original elements in the universe. Fast-forward two billion years, and vast gravitational forces have slowed the expanding universe down, causing large quantities of hydrogen and helium to coalesce into what eventually became galaxies and stars. Some of these stars were so hot they "burned" their hydrogen and helium into heavier elements, including carbon, oxygen, and iron—a chemical process that can take "only" a few hundred million years. When these stars eventually exploded as supernovas, they ejected their heavy elements into space, to become raw material for the next generation of stars.

Recent research indicates that our sun came into existence as the result of a localized "Little Bang" nine billion years after the big one. Scientists believe that an exploding supernova created a dense cloud of gas and dust that eventually birthed the sun and the solar system, thanks again to the work of intense gravitational forces. Additional carbon arrived later, most likely when nearby red giant stars blew off their outer, carbon-rich atmospheres before collapsing and turning into white dwarfs. This carbon traveled through space mostly as stardust, riding interstellar winds, before coalescing over the eons into comets, asteroids, and small planets. This stardust carbon eventually found its way to Earth, and eventually into us.

There are, by the way, carbon stars, to go along with the much more prevalent helium-and-hydrogen-dominated stars, such as our sun. Carbon stars typically contain more carbon than oxygen, which can give them a red appearance. In contrast, our sun is an "ordinary" star, meaning that it is richer in oxygen than carbon, which accounts for its yellowish color.

Current thinking among scientists is that most of the carbon on Earth arrived in a rain of comets early in its history, not long after the planet cooled. The early solar system at this time was a chaotic place, with all sorts of interstellar objects whizzing around, including numerous comets. This meant that the chances of an object striking Earth were pretty good, cosmologically speaking. And not only did these early bombardments likely deliver most of Earth's carbon, but according to one theory more than half of the element may have arrived via one massive comet impact.

All of this sets the stage for perhaps the biggest question of all: how did life begin on Earth? In other words, how did chemistry (elements like carbon, iron, and silicon) become biology (DNA, RNA, replication, respiration, etc.)? At the beginning of Earth's history, there was no life. About five hundred million years later, however, according to the fossil evidence, life existed, in the form of simple cells. What happened in between? Chemists said that the chances of a DNA molecule arising by random events from nonbiological material are very, very small. At the same time, conventional scientific thinking said it was highly unlikely that a DNA molecule arrived on a comet because DNA can't survive for very long when exposed to solar radiation. It was a puzzle.

This latter view, however, changed dramatically in 2005 when NASA's Deep Impact probe scored a direct hit on comet Tempel 1, producing a plume of gas and dust far richer in carbon compounds than scientists had expected. Sometimes called "dirty snowballs," comets are composed of rock, dust, water ice, and frozen gases—including methane, carbon dioxide, carbon monoxide, and ammonia. As the Tempel 1 experiment revealed, however, they also contain a wide variety of organic compounds, including long-chain hydrocarbons.

The plot thickened in 2009, when NASA announced that it had found an amino acid called glycine, one of the fundamental chemical building blocks of life, in the dust of another comet named Wild 2. As a result, Carl Pilcher, who led the NASA Astrobiology Institute, said, "The discovery of glycine in a comet supports the idea that the fundamental building blocks of life are prevalent in space and strengthens the argument that life in the universe may be common rather than rare."[4]

The plot thickened again in 2011, when NASA announced there was also a reasonable chance that meteorites may have brought source materials for DNA to Earth. Using advanced mass spectrometry instruments, scientists scanned eleven carbonaceous asteroids for nucleobases, which are part of the building blocks for DNA and RNA, and discovered three that were common in asteroids but were rare or absent on Earth. "Finding nucleobase compounds not typically found in Earth's biochemistry strongly supports an extraterrestrial origin," said Jim Cleaves, one of the scientists.[5]

Then, in 2012, astronomers reported the detection of a sugar molecule in a distant star system for the first time. According to an announcement, this finding suggests that complex organic molecules can form in stellar systems prior to the formation of planets, eventually arriving on young planets early in their history—possibly including Earth.

Once here, however, how did proto-RNA/DNA spark life? How did biology arise from chemistry?

For a long time, the best guess for the origin of life on Earth was the "primordial soup" theory—an idea originated by none other than Charles Darwin. In an 1871 letter to his friend and fellow scientist Joseph Hooker, Darwin speculated that life might have begun in a "warm little pond, with all sorts of ammonia and phosphoric salts, light, heat, electricity, etc. present, so that a protein compound was chemically formed ready to undergo still more complex changes."[6]

Darwin's theory lay dormant until the 1920s, when two researchers, Alexander Oparin and J. B. S. Haldane, working independently of one another, asserted that Earth's original oceans were a vast "primeval soup" of nonorganic molecules that bubbled and stewed for millennia, absorbing the energy of sunlight, until they "grew" organic molecules that could survive and reproduce on their own in hostile environments—as some bacteria molecules do today in hot springs and volcanic vents. Oparin argued that it only had to happen once. Life, once started, would take care of itself. The first living organism, both scientists said, would be little more than a few chemical reactions encased in a thin membrane to keep them from being destroyed. These organisms would grow by absorbing organic molecules around them, divide, and grow again. Eventually, photosynthesis would arise, and the oxygen it created would change Earth's atmosphere, making it amenable to other life-forms. Haldane called this process *biopoiesis*—a name that didn't catch on. His description of the oceans as a "hot dilute soup" did, however.

In 1952, this powerful metaphor received a significant jolt, literally. University of Chicago graduate student Stanley Miller and his professor, Harold Urey, decided to test the Oparin-Haldane hypothesis in what became one of the classic experiments of post–World War II science. Their goal was to re-create the prebiotic conditions of Earth's early oceans and atmosphere in the laboratory to see if they could generate organic compounds from inorganic ones. Speculating that volcanic activity would have released methane, hydrogen, and ammonia into Earth's proto-atmosphere, they sealed these gases in glass piping built in a closed loop. On one end of the loop was a flask filled with water, which was boiled to create water vapor; on the other side were two electrodes, representing lightning. After sparking the vaporous mixture with the electrodes, they cooled the gases and allowed them to "stew" for a few weeks before analyzing.

What they discovered made headlines around the world—and still forms the foundation of most scientific inquiries into the origin of life on Earth.

Miller and Urey discovered that as much as 10 to 15 percent of the carbon in the system (originating as methane) had formed simple organic compounds, and 2 percent had actually become amino acids—essential to life. In an interview at the time, Miller said: "Just turning on the spark in a basic pre-biotic experiment will yield 11 out of 20 amino acids."[7] More remarkably, in 2007 scientists reanalyzed the sealed vials from the original experiment, discovering that there were actually *more than twenty* different amino acids in the original mixture. In this way, the experiment strongly supported the Oparin-Haldane "primordial soup" theory, showing that simple organic compounds could be formed from gases with the addition of energy. Lightning, their experiment suggested, had provided the original spark of life on Earth.

Recent research has challenged parts of their conclusion, however. Investigations into the actual composition of Earth's atmosphere during its proto-development phase, called the Hadean Period (after Hades, the Greek god of the underworld), reveal that its chemical composition was more complicated than Miller and Urey had envisioned, including the presence of oxygen, which would have been hostile to the formation of organic compounds. While complicated, the picture emerging is one of an extremely turbulent, mostly liquid planet subjected to intense ultraviolet radiation, massive undersea volcanic eruptions, and frequent bombardment by rocky debris from outer space. These impacts would have kicked up large amounts of steam, which eventually blanketed the entire planet with hot, smelly clouds. Rain—and lightning—would easily have followed. It is quite possible, under this scenario, that additional amino acids arrived on Earth hitched to meteorites and comets and were tossed into the bubbling primordial soup like cosmic potatoes or carrots. At this point events likely followed the now familiar directions for life recipe: make soup stock; add carbon, heat, and energy; and let stew for a few hundred thousand years!

However, this raised another question: is life possible without carbon?

"Yes," said a group of scientists in a report published in 2007 by the National Research Council. They called it "weird life"—life with an alternate biochemistry from what is found on Earth. According to the report's authors, the fundamental requirements for life as we know it—water-based biosolvents, a carbon-based molecular system capable of evolution, and the ability to exchange energy with the environment—are not the only ways to support phenomena recognized as life. "Our investigation made clear that life is possible in forms different than those on Earth," said lead author John Baross, professor of oceanography at the University of Washington. But we'll never recognize it, he continued, if we're only searching for Earth-like life in outer space.[8]

"No discovery that we can make in our exploration of the solar system would have greater impact on our view of our position in the cosmos, or be more inspiring, than the discovery of an alien life form, even a primitive one," wrote the report's authors. "At the same time, it is clear that nothing would be more tragic in the American exploration of space than to encounter alien life without recognizing it."

The astronomer Carl Sagan once referred to this situation as "carbon chauvinism," arguing that life could alternatively be based on silicon or germanium. This may have been the inspiration for a famous *Star Trek* episode in which Captain Kirk and crew explore a planet dominated by aggressive and gooey silicon-based life-forms called Horta (an encounter with which prompts a memorable mind-meld with Spock). The trouble with silicon, however, is its powerful attraction to oxygen. Life, as we define it, requires a respiratory process, which removes waste. In carbon-based life-forms, the waste product is a gas, carbon dioxide, which is easily dispatched. The waste product of silicon, however, is sand—a solid. This means, according to biochemists, that it would be very difficult for silicon to provide a basis for viable life, even weird life.

As for the report's authors, they point to ammonia and formaldehyde as possible biosolvents that could support a home for weird life. They also noted that recent experiments demonstrate that DNA could be constructed from nucleotides based on sodium hydroxide and hydrochloric acid—meaning that an organism could have an entirely noncarbon-based metabolism. Weird life might even exist on Earth, they argued. We've just not tuned our minds to the possibility. Field researchers should therefore seek out organisms with novel biochemistries to better understand how life on Earth truly operates. This improved understanding will help us in our restless search for life beyond the confines of our blue-green planet.

Which suggests that all manner of coexistences might be possible if we're willing to look in the right places.

Which brings me to Dubbo.

Shortly after my visit with Eric at Gilgai, I spoke at the annual carbon farming conference, held in Dubbo, New South Wales. This event was originally organized by Christine Jones in 2007 as a way to build bridges between researchers, practitioners, and policy types around the idea of improving the carbon content of Australia's soils. It's grown into a sizable two-day event, as I discovered, with presentations on topics as diverse as cattle, photosynthesis, perenniality, sheep, roots, rhizomes, carbon pathways, wool, microbes, humus, glomalin, the soil food web, worms, birds, building topsoil, better water cycle management, niche marketing, resilience, profit, biodiversity, and something called "extreme carbon farming"—which dared to ask the question "How fast can you grow carbon on *your* land?"

Exhibitors packed the conference trade floor, and the agenda brimmed with researchers, farmers, and government representatives. The best line of the conference came from a farmer who said he had changed his stocking rate (normally meaning acres per cow) from 40 to 1 to 5 trillion to 1! He meant, of course, 5 trillion microbes to 1 teaspoon of soil. The best fact I heard was that 15 jackrabbits equaled 1 cow for grazing pressure (the public often forgets that grazers include rabbits, birds, grasshoppers, and other insects). The best quote of the conference came from the great American conservationist Aldo Leopold, via another Aussie farmer, who once observed, "Land is a fountain of energy flowing through soil, plants, and animals."

That certainly described what I saw Down Under, where all sorts of positive energy flowed through the farms I visited, from the grass to the animals to the people and beyond. I was especially impressed by the positive language Aussies use in day-to-day conversation, which is replete with "G'days" and "Good on yas" and

"Right-os." I suspect this can-do attitude lies at the heart of t innovations taking place today in Australia's sustainable agriculture community—that and the famous Aussie disdain for authority. Of course, the beer helps too.

It was a quote by the French contrarian Voltaire, however, that really set me to thinking: "Men argue, nature acts."

This quote opened a one-day workshop on carbon farming prior to the conference. The workshop was taught by Michael Kiely, one of the cofounders of the Carbon Farmers of Australia, a group that has been helping farmers gain access to carbon markets. Kiely used the quote to make a point about climate change and the completely inadequate efforts so far globally to either confront business-as-usual greenhouse gas emission production or prepare ourselves for the consequences of a changing climate. There's a whole lot of arguing going on, Kiely pointed out, and very little action. Meanwhile, nature is on the move, waiting for no one. Voltaire was referring to the human habit of arguing philosophical ideas to death, but his words certainly work as an analogy for our modern predicament. Aussies, by the way, are much more comfortable with the words "climate change" than Americans are. It's not the *crisis-that-shall-not-be-named* Down Under, thankfully—which is another reason Australia is leading the way on these issues.

In fact, Australia is *acting*. Two laws recently passed by the Australian Parliament and signed by Prime Minister Julia Gillard were hot topics at the conference and the subject of the carbon workshop. One implemented the world's first substantive, nationwide carbon tax, which placed a rising fee on the sources of greenhouse gases, such as power plants and coal mines. The idea is to drive down carbon emissions by driving up the price of pollution. It happened over the strenuous objection of the minority Liberal (i.e., conservative) Party, however, which has vowed not only to repeal the tax, but to bring down the Gillard government over it, which didn't sound very much like coexistence to me. In fact, the Liberals succeeded. In September 2013, the Liberal Party swept to power in national elections and is currently working vigorously to eliminate the carbon tax, as promised.

The second bill was called the Carbon Farming Initiative. Unlike the carbon tax law, it passed with broad support across the political spectrum. The bill creates a cap-and-trade system that everyone hopes will stimulate a carbon market, which, theoretically, will pay farmers for sequestering carbon dioxide in their soils and plants. The idea is to create a carbon marketplace where carbon credits can be bought or traded to offset the carbon emissions of a polluting entity. This marketplace requires a cap on total carbon emissions, whether regionally or nationally, so that a value or price can be placed on the credits themselves. In other words, if a polluter exceeds its cap by a certain percentage, then it can buy or trade for an offset that brings it back into compliance. Farmers benefit because they earn credits, to be sold, for their CO_2-sequestering activities on their land.

That's the theory, anyway.

Offsets sound like a good idea, but I began to have my doubts when the US Congress briefly considered its own cap-and-trade bill, way back in 2009, and word leaked out that the biggest player itching to get into the carbon marketplace was Goldman Sachs, the giant Wall Street investment banking firm that was so deeply involved in the colossal financial shell game that triggered the near-collapse of the global economy in 2008. What did Goldman Sachs want with carbon sequestration in farm and ranchland soils? Not to help the world avoid catastrophic climate change, I suspected. They wanted money, and lots of it. But in a carbon market? Yes, apparently. The middlemen, called *aggregators*, in a trading arrangement between a polluter and an offsetter (such as a farmer) earn a fee as part of the deal—a potentially big fee. It comes largely at the expense of the offsetter, too, which means farmers and ranchers won't earn as much money as they might expect. It all sounded a bit dodgy to me. In Australia, cap-and-trade-style plans are called *schemes*, and from what I could tell, Australia's scheme didn't protect its farmers any better than the US Congress's bill would have done.

It was a feeling reinforced at the Carbon Cocky Awards ceremony on the final evening of the conference. "Cocky" is Australian slang for "innovator," and the ceremony honored a half-dozen carbon farmers from across the nation in what was a lively, upbeat, and hopeful celebration of good work and camaraderie. I was impressed by the amount of carbon farming going on in Australia. However, I was also impressed by the two men in expensive suits who sat at our table during the ceremony. They were not farmers. In fact, they looked like sharks. One handed me his card, which identified him as an aggregator from Sydney. He smiled and shook my hand. He asked, could we talk later? When I told him I was a just a visiting American, the light went out of his eyes. Not interested. The shark swam on.

"This isn't the sort of coexistence we need," I thought to myself. "Predator and prey stuff. We have too much of that already."

Economically, we need something else.

What would that be exactly? I wasn't sure, but my trip Down Under made one thing very clear: thanks to the work of Colin Seis, Eric Harvey, Christine Jones, and a whole bunch of other cocky farmers and their allies, we have practices that can revitalize carbon in the soil. We also are getting a good idea of how to intensify food production regeneratively, which means we might have a good shot at feeding nine billion people by 2050. We know how to do these things. However, making them work at scale, which means making them work as part of the national economy, is the next big job. Good on the Aussies for trying. Pasture cropping and flerds are certainly innovations worth sharing. Carbon is key, but so are "can-do" attitudes, which Aussies have in abundance. With luck, what's happening Down Under will spread around the planet.

I'm keeping my fingers crossed.

A Carbon Sweet Spot

TWITCHELL ISLAND, SACRAMENTO–SAN JOAQUIN RIVER DELTA, CALIFORNIA

For a minute, I thought I had stepped into that scene from *Lawrence of Arabia* where Peter O'Toole, approaching the Suez Canal on foot, sees a ship sailing across the sand.

I had parked on an earthen levee at the eastern end of Twitchell Island, in the great Sacramento–San Joaquin River Delta, east of San Francisco. In front of me was prime farmland, and in the distance, just beyond a slight rise, I saw a big cargo tanker plowing its way slowly across a field à la *Lawrence*—plowing the middle of the San Joaquin River, of course.

I didn't drive to Twitchell Island to see farmland, however. I wanted to see a carbon *sweet spot* in action. Sweet spots are where big things happen in small places for a minimum amount of effort and cost. On Twitchell, a whole suite of big things had happened on just 14 acres of wetland in only a few years. Thanks to a high density of plant matter and a low rate of decomposition, wetlands are the world's best ecosystems for capturing and storing the carbon from CO_2 in their soils. Their destruction, conversely, releases lots of CO_2 into the atmosphere as these soils dry out and oxidize. Moreover, at least one-third of the world's wetlands are composed of peat, a type of soil created by dead or dying plants that are permanently water-bound. Peatlands, which include bogs and fens, contain 30 percent of global terrestrial carbon but cover only 3 percent of the earth's land surface (8 percent in the United States)—which is a lot of carbon bang for the buck.

Alas, of the approximately 200 million acres of wetlands that existed in the United States during the 1600s, more than half have been destroyed, mostly by draining and conversion to farmland or commercial and residential development. Although the rate of destruction has slowed considerably in recent years thanks to our understanding of the critical role wetlands play in ecosystem health, roughly 60,000 acres are still being lost annually.

The Sacramento–San Joaquin River Delta was once a vast freshwater marsh thick with tule reeds, cattails, and abundant wildlife. At least six thousand years old, the marsh caught sediment that washed down annually from the Sierra Nevada range, building up soil that eventually reached 60 feet deep in places. When the delta began to be settled in the 1860s following California's famous Gold Rush, farmers couldn't believe their luck. Since the soil had often been submerged—a consequence of flat terrain, frequent flooding, and tidal action—it had essentially become peat, rich in carbon and other organic minerals. Crops grew vigorously in the fertile soil. Soon a new kind of gold rush was on—to grab land in the delta, drain it, and grow row crops by the bushel-load. By the 1930s, this "reclamation" work, as the government described it, was largely complete. Over 1,150 square miles of former wetland had been diked into fifty-seven separate islands, each surrounded by a levee.

That's when the land began to sink.

Today, 98 percent of the delta lies below sea level, with many of the islands 10 to 25 feet down, requiring pumps to work continuously to keep plant roots dry enough to grow the crops. This sinking is called *subsidence,* and it starts when organic carbon in peat soils is exposed to air. Waterlogged wetland soils are *anaerobic*, or oxygen-deprived, which means organic carbon accumulates faster (via annual plant growth) than it can decompose. However, drainage in the delta created *aerobic*, or oxygen-rich, conditions in the soil, which encouraged rapid microbial digestion of the carbon, released into the atmosphere as carbon dioxide. In other words, the carbon literally vaporizes. As a consequence, the rest of the peat dries up and either blows or washes away. Scientists have calculated that the rate of soil loss in the delta under these conditions can be as much as 1 to 3 inches *per year*.

These lowering soil levels are putting a great deal of stress (hydrostatic pressure) on the levees requiring that they be continually raised and strengthened to prevent their collapse, a costly and time-consuming business. And ongoing subsidence creates the potential for a catastrophic failure of the levee system as the result of a major flood or earthquake. Breaches in the levees, though rare, have caused serious problems in the past. As if that weren't enough, subsidence encourages salt intrusion from San Francisco Bay.

It's not just row crops that are threatened, however. Two-thirds of all Californians get some part of their drinking water from the delta, and much of the state's agriculture depends on the delta for a steady supply of (salt-free) irrigation water.

Not many people know that central California is a mammoth plumbing project, crisscrossed by a complex network of canals, ditches, and pumping stations. And most of the water in this plumbing system originates in the Sacramento–San Joaquin River Delta. Originating in watersheds that encompass 35 percent of the state of California, nearly half of all the state's total river flow winds up in the delta, of which 7.5 million acre-feet of water, or roughly 25 percent, is delivered to huge pumping stations in the southern delta. Over 80 percent of this water is delivered to agriculture, supporting a nearly $45 billion industry in the state. Much depends on maintaining the integrity of this vast plumbing project. That's why subsidence, weakening levees, expanding salinity, and rising sea levels due to climate change all combine to keep California's water managers up at night.

Enter a group of scientists with the US Geological Survey in Sacramento, led by Robin Miller. In 1997, she and her colleagues came up with a novel idea to reverse subsidence in the delta: resurrect the marsh. Could the process that created the land-loss problem be reversed, they asked? In other words, could controlled flooding re-create the original marsh ecology and thus begin to build the soil back up as it did before?

To find out, they implemented an experiment on two 7-acre, side-by-side plots of farmland adjacent to a ditch that bisected Twitchell Island, located in the northwestern part of the delta. Twitchell Island had been "created" in 1869 by ditches and levees, and by 1997 it had dropped 6 meters (18 feet) below sea level. The scientists determined that 3 to 9 meters of peat soil remained on the island, which meant further subsidence was likely.

To test their hypothesis, they flooded the western 7-acre plot to a depth of 10 inches and the eastern plot to 22 inches. Tules were planted in a small portion of both plots. By the end of the first growing season, cattails had colonized both plots (the seeds arriving on the wind), which provided a screen for other plants, including duckweed and mosquito fern. Then things really took off. After just a few short years of annual managed flooding, the western plot had developed a dense canopy of marsh plants, as did the eastern plot, though it maintained some open water.

When they took measurements of the soil after seven years, they were amazed to discover that the soil in both plots had risen *10 inches*—the result of 15 tons of plant material growing and dying per acre per year.

This was good news.

"Ten years after flooding," wrote Miller in a peer-reviewed summation, "elevation gains from organic matter accumulation in areas of emergent marsh vegetation ranged from 30 to 60 centimeters [1–2 feet], with an annual carbon storage rate approximating 1 kg/m^2, while areas without emergent vegetation cover showed no significant change in elevation."[1]

This answered their research question: subsidence could be reversed by returning natural marsh processes to the land.

But the hopeful news was just beginning. The researchers tested the amount of CO_2 that had been sequestered in this new soil as a result of their experiment. They suspected that 1 to 2 feet of dense, carbon-rich peat soil likely soaked up a lot of atmospheric CO_2—and they were right. In fact, as much as 25 metric tons per acre per year were sequestered in the study plots, according to their analysis. In comparison, a typical passenger vehicle emits 5 metric tons of CO_2 per year. The 14 acres in the study plots sequestered the equivalent emissions of seventy passenger vehicles per year!

According to the scientists, if the entire delta could be turned into carbon farms, the net CO_2 effect would be like changing all of California's SUVs into small hybrids or the equivalent of turning off all residential air conditioners for a year. And that doesn't even count the CO_2 emissions eliminated by not farming the land (conventionally, anyway). *And* that doesn't count all the other ecosystem services generated by a functioning marsh, including water purification and wildlife habitat.

The researchers called what they did a "carbon-capture farm" and hoped that the project would demonstrate that it is highly feasible to use managed wetlands to sequester carbon and reduce subsidence simultaneously. Although the specifics of this project are likely limited to the Sacramento–San Joaquin River Delta, it is nonetheless a very good example of a sweet spot. On just 14 acres, the project demonstrated how to (1) reverse subsidence, (2) reduce the risk of levee failure, (3) sequester a lot of carbon, and (4) provide wildlife habitat, especially for birds on the Pacific flyway.

The experiment also raised other questions, which the team hoped to answer with a scaled-up project somewhere else in the delta: Is there some way to increase carbon storage in the soil beyond current levels? Could this be a buffer against rising sea levels? What are the total greenhouse gas emissions produced by restored wetlands, including methane and nitrous oxide, and how will this effect the overall goal of net greenhouse gas reduction?

Unfortunately, the answers won't be available anytime soon. That's because the project was cancelled.

"The State of California stopped supporting the research when the budget problems hit," Miller wrote to me in an e-mail. "Still, while the project was active, much valuable information was gained."

One question that wasn't addressed in the research occurred to me as I watched the tanker sail across the farmland on Twitchell Island that day: where's the economic incentive to "carbon-farm" this land? For all the multiple benefits that restoring the marsh ecology would create, there's no incentive for a private owner to stop farming here—not yet.

The scientists assumed the answer lay with the creation of a carbon market via a cap-and-trade program, in this case via a law called the California Global Warming Solutions Act of 2006, signed into law by Governor Schwarzenegger, with the goal of reducing

California's greenhouse gas emissions to 1990 levels by 2020. Similar to Australia's model, the law required California to place a limit (or "cap") on major emitters of greenhouse gases, such as power plants, fuel refineries, and the transportation sector. These caps will then decline by 3 percent each year. Concurrently, the state will distribute *allowances,* which are tradable credits, equal to the emissions allowed under the cap. Emitters over the cap must buy credits or otherwise offset their emissions or else face financial penalties. Ideally, market forces will spur technological innovation and investments in clean energy so that emitters can come under the cap and thus sell their allowances.

The farmers on Twitchell Island could benefit by selling offsets they had generated by converting their farmland back to delta marsh and sequestering all that lovely carbon! The linchpin is what scientists call *additionality*—meaning the carbon added to the soil is new or additive to the carbon already present. It's this extra carbon that forms the basis of a tradable carbon credit. The key is hard numbers, which Miller and her colleagues had in spades. All that Twitchell Island farmers needed was a marketplace. Right?

Maybe. Many observers remain skeptical of the offset concept. When I first looked into it, I discovered a serious objection: "cap-and-trade" can easily become "cap-and-swindle." At best, offsets may be illusory; at worst they may be fraudulent, thus imperiling the whole idea.

However, there was nothing illusory about what had happened on Twitchell Island, despite my *Lawrence of Arabia* moment. That's why some are touting the demonstration project as a role model for other work, including Stephen Crooks, a San Francisco–based wetland restoration expert. "This is probably the highest sequestration of carbon dioxide you can get in a biological system," he said in an article. "This is the foremost example of showing how you can restore wetlands and sequester carbon at the same time. We can use what has been learned as a very firm reference to help inform policy development in the United States and overseas."[2]

If a market could be created for wetland restoration, Crooks went on to say, it would have multiple benefits beyond carbon sequestration, including recreational opportunities for hunting, fishing, and birding created by the restored wildlife habitat, which, incidentally, would be an additional source of income for landowners. In fact, the Twitchell Island project demonstrates that all sorts of possibilities exist when we adjust our lenses to look at the world in a different light.

Starting with just a few acres at a time.

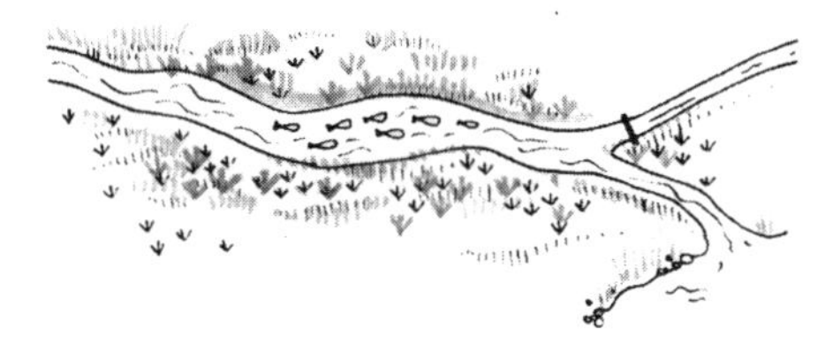

4

RESILIENCE

It was an odd feeling to stand on dry land 6 feet under sea level and think everything was just dandy.

That's what I felt after arriving in New Orleans on a rainy August afternoon in 2012. After checking into my hotel, I went for a walk along Bourbon Street, knowing that the city's famous French Quarter sat in a shallow bowl squeezed between the Mississippi River and Lake Pontchartrain—a bowl whose uneven bottom reached 6 feet below sea level in places. Almost anywhere else in the world that would mean a snorkel and swimming trunks, but New Orleans isn't like many other places, that has some people very worried. I had read all about it in a report that analyzed the city's chances of survival under climate change and rising sea levels, rating them as poor. It wasn't simply an issue of depth—the bowl's top is dikes and levees, which will have to be elevated and strengthened to meet rising sea levels. They could also buckle again under the stress of an intense storm surge, as Hurricane Katrina notoriously demonstrated in September 2005. This one-two punch made the future of the Big Easy perilous at best.

You would not have known it, however, by walking through the streets of the city. On a long ramble, I saw life-as-normal everywhere I went—if you can call what takes place in the French Quarter "normal." Energetic horn and drum bands performed on busy street corners as tourists in T-shirts strolled in knots, beer cups in hand. Unsmiling hawkers urged passersby inside darkened doors as hooting, clean-cut young men on balconies urged equally clean-cut young women below them to flash certain body parts. Large signs outside bars advertised an ominous drink called a "hand grenade," while families with children sauntered incautiously along the streets. Music poured from open windows, and everywhere strings of cheap beads dangled from protuberances or lay guttered in the street. Their sheer quantity was astounding. I even discovered a string of party beads around the neck of Benjamin Franklin in a park; he looked like he had been up all night drinking with the locals before sneaking back to his pedestal.

For a city possibly without a future, there sure was a lot of partying going on—and the sun hadn't even set yet!

It's not like New Orleans isn't aware of its precarious existence, however. All you had to do was read the headlines in the *Times-Picayune*, as I did upon my arrival, or turn on the TV. A historic drought in the Midwest had sent the Mississippi River to record low levels, causing all sorts of consternation. Saltwater from the Gulf of Mexico had reached Chalmette, only 4 miles downstream from where I rambled, forcing the town to declare a state of emergency and beg New Orleans to send it millions of gallons of drinking water. New Orleans itself was next in line for trouble. The saltwater "could take out the water supply for all of us if we're not careful," said Deputy Mayor Cedric Grant after Chalmette's emergency request was approved. Meanwhile, the US Army Corps of Engineers promised to construct a 1,700-foot-long underwater dam on the Mississippi below Chalmette in order to block the salty, denser Gulf water from moving farther upriver—at a cost of $5.8 million. Not coincidentaly, it chose a location where similar dams had been built in 1988 and 1999. Apparently, this situation had happened before.

It's not easy being 6 feet under, I began to see.

Twelve months earlier, the crisis was 100 percent flipped. Historic floods on the Mississippi River had forced evacuations up and down its length, causing billions of dollars in economic damage to the nation. These floods also moved a great deal of soil (mostly from farms) into the river, elevating its bed. Then came the historic drought. According to the federal government, the month before my visit was the hottest July on record for the United States. Water levels in the Mississippi were 50 feet lower than the previous year, and in some places there was only 5 feet of water left in the river, causing commercial barges to run aground. A representative of the American Waterways Operators called the situation "near critical," noting that 60 percent of the nation's grain, 22 percent of its oil, and 20 percent of its coal move up and down the Mississippi each year. It would take sixty tractor-trailer trucks to carry the cargo contained in just one barge, he said, and 144 eighteen-wheeled tankers to carry the oil from one petroleum barge.

"We've just dealt with a historic flood, then the water drops," said one Louisiana official. "We have some fifty-year guys who've never seen anything like this before. It's a completely different river than anybody's ever seen."

Maintaining the human hold on the mighty Mississippi, it seemed clear, had become a nonstop ride on a bucking bronco of anxiety and expense.

Easing both was one reason I had made the trek to New Orleans—to explore an idea being pushed by a group of innovators to employ nature and the marketplace to improve the city's chances of survival. It involves an ambitious plan to restore the Mississippi Delta to ecological health, including resurrecting its formerly extensive coastal wetlands, which would buffer the city from surging seas churned up by future hurricanes. There is very little doubt among scientists that sooner or later another superstorm will strike the city, juiced by climate change as Katrina had been. Is New Orleans resilient enough? It had bounced back from Katrina, but just barely and only with a great deal of expense and suffering. Could it bounce back again from another devastating blow? What about rising sea levels? According to scientific projections, sea level will rise 1 to 6 feet by the end of the century. Already 6 feet under, how does New Orleans remain resilient in the face of this threat?

What were the options, I wondered, besides partying endlessly?

One answer involved something called *blue carbon*, which is a term used by scientists for the carbon stored in the soils of coastal wetlands, cypress swamps, mangrove forests, and seagrass beds. The "blue" part is a metaphorical flourish, signifying the carbon's relationship to the sea. Living at 7,000 feet above sea level, I had to admit I had never heard the term before. I had heard of black carbon, which is the sooty stuff created by woodstoves, forest fires, diesel engines, and certain types of dirty power plants, as well as green carbon, which is what researchers call the carbon locked up in the earth's vast temperate and tropical forests. I had even heard of red carbon, which is a type of cool-looking vinyl used in high-end commercial products, including car seats and electric guitars. And I suppose you could describe the glomalin-rich humus in gardens, farms, and ranches as brown carbon . . . but blue carbon? What did it do? And why did researchers insist that it could play a big role in strengthening New Orleans's resilience?

I decided to ask Sarah Mack, one of the leaders of the effort to restore the coastal wetlands of the Mississippi Delta. A few months earlier she and her private company, Tierra Resources, made headlines for authoring the first protocol in the nation that proposed to use market-based carbon offsets to restore coastal wetlands. This methodology has the potential, according to a news story I read, to pump $5 billion to $15 billion into Louisiana over the next four decades while mitigating climate change and shoring up coastal defenses against hurricanes. It was an effort that took Mack five years and netted her the $50,000 top prize in the 2012 Water Challenge, sponsored by the Greater New Orleans Foundation for up-and-coming entrepreneurs. If that wasn't impressive enough, she was also the first scientist I had ever met who sported a visible tattoo—a function either of her youth, her New Orleans environment, or my relatively sheltered life.

Probably all three, I decided.

To say her job is daunting only scratches the surface of the challenge. It starts with the Mississippi River and its "taming" over the past century in one of the greatest feats of engineering in human history. Nearly 2,400 miles long from its origin at Lake Itasca in northern Minnesota to its mouth 100 miles below New Orleans, the Mississippi flows through or borders upon ten states and receives tributary waters from twenty-one more, plus two Canadian provinces. At 1,250,000 square miles, its watershed is the fourth largest on the planet and encompasses approximately 40 percent of the surface area of the contiguous United States, supplying drinking water to twenty million Americans daily. The river's annual water discharge into the Gulf of Mexico is roughly 600,000 cubic feet per second (4.5 million gallons per second), which, though puny compared to the Amazon River's 7 million cubic feet per second, makes it the largest of any river in North America by far. It is also a big reason why the Mississippi's story is inextricably bound up with floods and silt.

Geologically, the river is a product of the great Laurentide Ice Sheet, which twenty thousand years ago covered a large portion of North America. As the ice sheet retreated, rivers grew and hundreds of feet of rich alluvial soil were deposited across the Midwest by their floods, where they eventually became the fertile farm fields that we know today. Scientists estimate that the amount of sediment carried down the Mississippi to the sea prior to 1900 was about 400 million metric tons per year (1 metric ton equals 2,200 pounds), which replenished the vast wetlands and bayous of its delta with fresh silt. Today less than 150 million metric tons make it to the coast per year—sadly so. The difference is a direct consequence of our engineering prowess. There are forty-three dams on the Mississippi alone and thousands more on its tributaries, each trapping vital sediment. Additionally, the river has had meanders cut off, rapids bypassed, its bottom dredged, and its banks armored with dikes and levees all along its length—all in the name of flood control, navigability, and commerce. The Mississippi has essentially been put in a cement and steel straitjacket and told to behave itself.

Now we're having big regrets.

What we've done to the Mississippi is a useful analogue for the way we've treated the planet in general. Project by project, mile by mile, we arm-wrestled a significant and magnificent natural system into crass servitude without considering the long-term implications of our actions. As a consequence, we're in a pickle of large proportions.

A good example is what we've done to the delta's vast wetlands and bayous (a type of really wet wetland). According to researchers, in only seven decades over 2,000 square miles of wetlands—an area larger than Delaware—have vanished due to the loss of replenishing sediment and freshwater from periodic floods. Louisiana's wetlands are being lost today at a rate of one football field *every hour*. They're being drowned, depleted of freshwater, pierced for oil and gas exploration, and battered by storm surges. This is bad news, not only because the region's wetlands, swamps, bayous, and barrier islands comprise one of the planet's most diverse ecosystems, but because they also contribute billions of dollars to the economy via seafood production, tourism, and port services. All in jeopardy. Toss in both the devastating effects of Hurricane Katrina, which essentially stripped Louisiana of its outer ring of shoreline defenses, and rising sea levels due to global warming, and you have the recipe for serious trouble.

What to do? Fortunately, Mack and her colleagues Rob Lane and John Day, professors of coastal ecology at Louisiana State University in Baton Rouge, had an answer.

Mack and Day picked me up near my hotel the morning after my arrival. Under gray skies, we headed out in Day's well-traveled car southeast toward St. Bernard Parish and two wetland restoration projects they wanted to show me. An intense, rumpled professor type, Day confirmed reports that a large amount of coastal land around New Orleans will be underwater by 2100 thanks to rising sea levels and declining rates of sediment deposited by the Mississippi River.

"Actually, it's worse than published forecasts," he said as we navigated down a busy street. "Almost everyone is lowballing the amount of land that is going to be lost. The latest numbers show that land is sinking faster than predicted and sea level is going to be higher. But everyone is living in denial."

As an illustration, he pointed out the car window toward a shiny-new big box store, part of a national chain, which had been ruined by Katrina and rebuilt. "It was a waste of money," he said bluntly. "New regulations required that it be constructed on slightly higher ground to protect it from future flooding. But they're just kidding themselves. This place is going to be under the sea if we don't change things fundamentally."

Looking at the building, I noticed that the parking lot was nearly full. I tried to imagine the cars sloshing through seawater as they crossed the lot, but couldn't. I got his point, however: tweaks won't cut it. Rebuilding on higher ground doesn't change the calculus in the long run. What's needed is something much more transformational.

If southern Louisiana is serious about its future, Day continued, it has only one viable option: restore its wetlands, swamps, and bayous with river sediment and green plants. The quickest way to accomplish this would be to let the Mississippi River flood again, over and over and at just the right time in order to maximize the limited amount of sediment available in the river. *And* a flood has to happen at just the right place with just enough velocity to spread the sediment out over the swamps and wetlands.

"Doesn't sound easy," I replied, earning a chuckle from the professor.

"Damn near impossible," he said. "Ecologically, politically, economically, culturally—you name it. That's why we're trying to restore wetlands by other means. Either way, there's really no choice, not if New Orleans wants a future."

"It's not like the city isn't trying," Mack chimed in. "It knows it has a big problem." There's a new master plan being developed that takes sea level rise into account, she said. The city is also supporting the US Army Corps of Engineers in its efforts to make the levees bigger and better. Many residential buildings are being elevated in vulnerable parts of the city, emergency services are being improved, water supplies are being reinforced, and a plan has been developed to address land subsidence within city limits, but it requires funding.

Much of this is Dutch-inspired, she said. A sizable portion of the Netherlands is below sea level, and after a devastating storm of its own decades ago, the Dutch government embarked on a comprehensive dike-building and flood abatement program designed to keep its citizens safe and dry. After Katrina, a number of civic leaders traveled to Holland to take a look. One idea they are considering is a "floating home" for the elderly and disabled—homes that essentially sit on an inflatable raft that can rise with floodwaters.

There was a pause in the conversation.

I suddenly realized why—we had driven straight into a very large puddle of water. Neither Day nor Mack seemed terribly concerned by this development, though I was. Inquiring, I was told that rainstorms frequently filled the lowest portions of New Orleans's "bowl" with water, which, because it was 6 feet under, had nowhere to go until the pumps finished their work. Apparently, that hadn't happened yet. I looked out the car window. The water was thick. As the car in front of us came to a stop, Day nonchalantly maneuvered around the vehicle, aiming for an elevated highway on-ramp ahead of us. I craned my neck, looking around, experiencing another odd feeling. I'm used to water running downhill, but when you live in the bottom of a bowl 6 feet below sea level, *there is no downhill.*

As we cleared the puddle, Mack picked up the conversation, saying there were many ways a city like New Orleans could manage its floods better, including innovative street designs, rain gardens, better drainage canals, and even artificial bayous and wetlands in the city. Things New Orleans *could do*, she said, but hadn't yet, even seven years after Katrina. She paused, letting the comment hang in the humid air.

"A river delta is a strange place to build a city," Day interjected. "Deltas sink, and when they're not sinking they get flooded." There have been twenty-seven major floods over the past two hundred years on the river, he said, repeatedly testing the resilience of city residents. That's not counting the yellow fever epidemics of the nineteenth century and the drinking water pollution problem early in the twentieth century. Then there are the hurricanes. Louisiana has been struck by 50 of the 275 named hurricanes that made landfall between 1850 and 2004, 18 of which were Category 3 or above. On average, a major hurricane comes within 100 miles of New Orleans every decade. Then, of course, there was Katrina—a behemoth that flooded 80 percent of the city, damaged 70 percent of its buildings, caused the death of over 1,800 people, forced 1.2 million people to evacuate (disastrously), displaced 100,000 people from their homes more or less permanently, and resulted in nearly $150 billion in damage.

A strange place to build a city indeed—and yet here it is.

"Is New Orleans doomed, in your opinion?" I asked Day as we traveled along the elevated highway.

It was a question he's been asked before, apparently.

"Science isn't about doom and gloom," he said quickly. "If you fall out of an airplane at 25,000 feet without a parachute you are going to die. That's just a fact. It's only gloomy if you didn't mean to jump."

A little while later, I thought about his words as we traveled through a neighborhood dotted with homes broken by Katrina and left listing in her wake like battered and abandoned dories. I stared in silence out the window. Why hadn't the buildings been torn down? Why were they left there to list and rot? It made the neighborhood look depressed—as in sad. My home state of New Mexico is one of the poorest states in the nation, but poverty back home didn't look like this, as if a small plague had swept through. As if nobody cared much about the past or the future. That couldn't be right. People *do care* about the past and the future, so something had happened to this neighborhood, something that didn't involve just gale-force winds, storm surge, and busted levees. It was something to do with people's spirits, as if they had fallen from Day's airplane—without meaning to jump.

My spirits perked up when we pulled off the road to examine one of the wetland restoration projects. Climbing out of the car, however, I was initially confused. We had parked near a small gray building with the words E. J. GORE PUMPING STATION emblazoned on its front. Round back was a shallow pit of water from which six large, serious black pipes emerged, only to disappear quickly under a thin layer of dirt and grass. They reappeared a short distance further at the edge of a swampy wetland, pointing at it like six giant fire hoses. I assumed they aimed not to put out a fire, but to fill the bayou with water like a bathtub. Why? I wasn't used to restoration projects that involved diesel pumps or large pipes, and as we wandered the freshly mowed earthen levee that separated the bayou from wherever the pump house got its water, I began to feel a bit bewildered by the complexity and strangeness of it all.

Fortunately, Mack was sympathetic to my plight. In her call-it-as-she-sees-it manner (possibly another reason for her tattoo), she explained that as a nonnative herself it had taken a while to adjust to life in the land of jazz bands and bayous.

She grew up on a farm along the Kansas-Colorado state line, where her parents practiced no-till farming, and where she inherited 800 acres that she manages at a distance today, she told me. In high school, she became curious about science, so she decided to major in chemistry at college. Carbon, however, wasn't on her mind yet. Instead, she focused on the toxic effects of agricultural chemicals—motivated, she admitted, by the fact that her brother had been born with Down syndrome. After graduation, she enrolled in Tulane University's graduate program in public health, which brought her to New Orleans. While earning a master's degree in toxic and hazardous waste management, she began an internship with the city's Sewerage and Water Board, which was founded in 1899 to combat disease, provide safe drinking water, and eliminate the health hazards of sewers—important jobs still today, apparently. In 2002, she accepted a job offer from the board, while continuing to work toward her Ph.D., in what she said was a fateful decision.

That's because three years later Hurricane Katrina slammed into New Orleans. After briefly evacuating the city, she returned as a primary emergency liaison, working with all agencies to keep the water pumps working, living day and night in a pump house like the one nearby, and putting in twenty-four-hour shifts as she and her fellow workers tried to keep New Orleans from drowning completely. All of the pumps failed or were down at some point, she said, adding to the sense of chaos that engulfed the Big Easy in the days following the hurricane.

"Martial law essentially ruled the city," she said, "and there were fires all over the place, despite the flooding." It made for a frightening and surreal experience. She ended up living for three weeks in that pump house. "It was an anything-goes atmosphere," she said, "and basically almost anything did."

Making matters worse, in late September New Orleans was battered by Hurricane Rita, a Category 5 monster that scientists later described as the fourth most intense hurricane ever recorded in the Gulf of Mexico. Crossing the same abnormally warm stretch of Gulf water that juiced Katrina a month earlier, Rita rapidly intensified as it approached the Louisiana coast, creating a panicky sense of déjà vu. Although Rita ultimately delivered only a glancing blow to New Orleans, its storm surge flooded low-lying communities and topped levees already battered and broken by Katrina.

Two devastating hurricanes in one month got Mack thinking about wetlands. Clearly, they needed to be restored, pronto—which gave her an idea. What if the city pumped its treated wastewater into nearby wetlands to provide the nutrients and freshwater formerly delivered by the Mississippi River, rather than simply pumping it into the river and out to sea, which was the current strategy? It might not provide the mineral sediments that the river once supplied, but the freshwater and the nutrients would stimulate wetland growth and result in organic soil formation and hence . . . carbon sequestration! It was worth a shot, she thought.

Inspired by the Clinton Climate Initiative, a program of the foundation that President Bill Clinton started after leaving the White House, Mack took a leap and wrote a grant based on her idea. She was elated when the grant was approved, but when she tried to implement it she ran into politics at her day job, as she put it, which stymied the project. Eventually, she felt frustrated and angry by the inaction, so when Day, who served on her Ph.D. committee, asked her to work on a Katrina-related lawsuit on behalf of flood victims, she quit her job and signed up.

It was another important turning point.

The object of the lawsuit was an infamous shipping lane, located less than 3 miles from where we stood, called the Mississippi River–Gulf Outlet canal, or MRGO ("Mister Go") for short. The nickname was apt. The 76-mile canal was dug in the 1960s by the Army Corps of Engineers, under orders from Congress, as a shortcut between the Port of New Orleans and the Gulf of Mexico, bypassing the time-consuming twists and turns of the Mississippi River below the city. It was part of that era's "go, go, go!" attitude toward commerce, abetted by our (supposedly) infallible engineering expertise and our disregard for the needs of the natural world. Almost as soon as MRGO was built, however, the introduction of lethal saltwater into freshwater wetlands triggered massive vegetation die-off and environmental degradation. The banks of MRGO began to erode (not being a natural waterway), tripling its width in only twenty years and causing shoaling underwater, which slowed barge traffic down. It also eviscerated vast stretches of cypress wetlands, turning them into open water as salt crept in from the Gulf, killing the vegetation. The end result was 60,000 acres of direct wetland loss and nearly 300,000 acres lost indirectly.

MRGO's real crime, however, was the role it played during Katrina's bombardment of New Orleans, where it earned another nickname—"Hurricane Highway"—for its amplifying effect on storm surge. Levees along MRGO ruptured in numerous places, as they did on the nearby Industrial Canal (itself an earlier attempt to shortcut the Mississippi), causing extensive flooding in adjacent St. Bernard Parish and other communities. It shouldn't have been a surprise. Three months earlier, a storm surge expert at Louisiana State University publicly called MRGO a "critical and fundamental flaw" in the Army Corps' hurricane defense system that could amplify storm surges between 20 and 40 percent. A post-storm investigation supported his analysis. The Army Corps of Engineers disputed this assessment, however, saying Katrina was essentially an act of God, and vigorously defended itself in the lawsuit that followed—the same lawsuit that Day asked Mack to join as an expert.

The Corps lost.

In November 2009, a federal judge ruled that the Corps—and thus the federal government—was liable for much of the damage caused by Hurricane Katrina. The landmark ruling awarded over $700,000 to four plaintiffs from the city's Lower Ninth Ward and St. Bernard Parish, which took the brunt of the flooding. "It is the court's opinion that the negligence of the Corps, in this instance by failing to maintain the MRGO properly, was not policy, but insouciance, myopia, and shortsightedness," the judge wrote. Meanwhile, the Corps had closed MRGO to all traffic—in its own way an admission of culpability. In an illuminating twist, a portion of the earthen dam that closed MRGO was built on top of a natural ridge with the express intention of facilitating wetland restoration efforts—including the one we were viewing at the moment. Ironically, the wetland in front of us was Mack's grant in action—using stormwater to fill a salty wetland, only without the office politics.

Unfortunately, the verdict in the MRGO lawsuit was reversed by the US Supreme Court in 2013, which decided that a previous ruling in a flooding lawsuit exempts the federal government from responsibility for flood damage. Nevertheless, there were two important consequences in the aftermath of Hurricane Katrina.

First, governments at various levels said they were now serious about wetlands restoration in the delta. In 2007, Congress approved a $284 million diversion project at Myrtle Grove, farther down the Mississippi, which aimed to replenish adjacent wetlands with freshwater as well as supply sediment mined from dredging projects elsewhere on the river, all with the goal of filling in open water areas and increasing plant productivity, thereby preventing future erosion and land loss. The project is part of a $50 billion, fifty-year master plan put together by Louisiana's Coastal Protection and Restoration Authority, itself a product of post-Katrina finger-pointing. Although Congress has not yet funded the Myrtle Grove diversion (alas), state officials have said they expect to spend $500 million a year on coastal restoration projects. For its part, the federal government issued a report in 2011 from President Obama's Gulf Coast Ecosystem Restoration Task Force, which recommended that the river's restoration needs be given the same weight as the more traditional goals of protecting navigation interests and flood control.

It's a good start, Mack said as we began to walk back to the car, but it's not enough land or money to make a difference. "If you fly over the delta in a helicopter, the areas being restored by these projects look like postage stamps. We've got to work at much bigger scales."

One challenge to this goal involves property rights. Approximately 80 percent of the delta's wetlands are privately owned, and even when the government and landowners agree on a restoration need—stopping erosion, for example—there is often spirited disagreement as to the proper course of action. Pumping freshwater into salty wetlands, for example, is often opposed by landowners who have built businesses around oyster, shrimp, and fishing enterprises that depend on salt water. It often adds up to inaction—even though everyone understands that action of some sort is required.

It all sounded like an impossibly tall mountain to climb. As we reached the car, I began to feel bewildered again. Fortunately, this is where the second positive consequence of Katrina comes in: Mack's decision to create a blue carbon market as a way to increase the resiliency of the region.

After completing her Ph.D. in global sustainable resource management at Tulane, she went to work on a scientifically rigorous methodology that could create a private carbon-based marketplace to do wetlands restoration on the kind of scale that would make a difference. Trouble was, nothing of the kind existed *anywhere in the world*. So, with a generous grant from the Entergy Corporation, the regional power company, through its Environmental Initiatives Fund, she enlisted the help of Day and Lane and began working out the technical details of blue carbon storage, verification, and management. The goal was to create a methodology that could be certified by a third (neutral) party as a "transactionable" carbon credit in the marketplace, as she put it.

The idea was straightforward: Wetland plants and soils are a phenomenal place to store carbon pulled from the atmosphere by photosynthesis. If they can be restored from a degraded condition to a healthy condition by a management action, then an amount of *new* carbon will be sequestered. Since this amount can be scientifically quantified, landowners can calculate the amount of carbon dioxide and other greenhouse gas emissions the restored wetlands will sequester over time. This creates a carbon credit, certified and registered with a third party, which can be sold to a polluting business, such as an oil refinery, to offset its greenhouse gas emissions. The proceeds from the sale of carbon credits would then help pay the costs for wetland restoration work.

"Our methodology gives you the recipe on how to create the project," Mack said, "how to quantify the carbon in a way that it can be transacted through an emissions trading market like any other commodity and bring in private money to invest in wetlands restoration."

According to her methodology, roughly 1 carbon credit = 1 ton of CO_2 (or CO_2 equivalent). About 4 million acres of wetlands are eligible for restoration in the delta, according to the master plan of the Coastal Protection and Restoration Authority, and Mack estimates that a healthy wetland sequesters as much as 15 metric tons of CO_2 per acre per year. At a market price of $12 to $25 per credit, $5 billion to $15 billion could be generated over forty years. Restored wetlands can result in numerous social and economic benefits, including the creation of jobs, an increased tax base, increased household income levels, and improved infrastructure. In this way, wetlands restoration can become a form of poverty alleviation. Private landowners like it because it's voluntary and keeps government largely out of the equation. Businesses like the market-based approach, scientists support it because it is empirically based, and conservationists like it because it gets the job done.

It looks like a win-win-win-win-win.

Mack said the methodology, published in a document called *Restoration of Degraded Deltaic Wetlands of the Mississippi Delta*, is not only the first of its kind in the world, it is the first to use a modular format, which means it's flexible and can be used for a variety of wetlands restoration techniques.[1] The methodology went through a tough eighteen-month peer-reviewed certification process before being approved in 2012 by the American Carbon Registry (ACR), a nonprofit organization founded in 1996 as the first private, voluntary greenhouse gas registry in the nation. Specializing in the development of high-quality carbon offset standards and protocols, ACR will approve the third-party verifier for the carbon credit marketplace in southern Louisiana.

According to Mack, the next step is to put all of the above to work on a pilot project in St. Charles Parish, located 20 miles west of

New Orleans, which began in the fall of 2013. Similar to the stormwater wetland restoration project Mack initiated behind the E. J. Gore pump house, a wastewater treatment plant in the parish will discharge treated municipal wastewater into an adjacent 950-acre wetland property to help restore the wetland's function and thus increase carbon sequestration. With the influx of nutrients, wetland grasses will grow, and Mack estimates that the project will sequester 1,000 to 7,000 metric tons of carbon dioxide annually. Her company, Tierra Resources, is planning to help "transact" the offsets in the next one to two years.

Another project involves planting mangrove trees in a salt marsh to see if their restoration can be a viable conservation strategy. This is important for two reasons: (1) mangrove swamps are among the most endangered marine habitats in the world, and (2) their extensive root system can anchor the trees against storm surge and keep wetland soils in place, allowing other plants to grow and improving overall wetland health and productivity. In addition to their role in protecting wetlands, mangrove trees also sequester relatively high amounts of carbon. The project's salt marsh, by the way, belongs to ConocoPhillips, the giant oil company, which owns over 600,000 acres of wetlands in the delta—not a bad partner to have if your goal is to think big.

There's an Achilles' heel to all this, however: *the carbon marketplace is voluntary*. Companies are not required to buy carbon credits by anybody, as they would be under a cap-and-trade compliance system like the ones in Australia and California. Instead, Mack and her colleagues are banking on the urgent need to restore the delta's wetlands as a motivator—so southern Louisiana can have a future. Is that enough, however, to compel a carbon marketplace into action? In 2012, the average carbon credit sold for less than $6 on the voluntary market in Louisiana, which was a 16 percent increase over the previous year, but still far below the $15 to $25 range that Mack had hoped to see. Still, she sounded optimistic. "Thanks to companies like Entergy, which have stepped up to purchase wetlands offsets voluntarily," she said, "we've established the foundation for a viable offset marketplace. It's a challenge and it's complicated, but I'm hopeful."

I'm hopeful too, despite my bewilderment.

Clearly, Mack and her colleagues have come up with an innovative idea, one that could possibly extend well beyond the Mississippi Delta. That's because the question of *resilience* in a rapidly changing world—one that promises to throw any number of big surprises at us—is looming larger and larger as the century moves along. If what we've done to the Mississippi River over the past century is an analogue for the pickle we're in globally, then some of the solutions proposed by Mack might point to a way out. I'm especially hopeful because restoring "ecosystem services" to health, as well as finding a way to pay for them, is the subject of a great deal of head-scratching around the planet right now.

Let's back up for a second and take a look at that term *ecosystem services*.

The term came into widespread use in 2005 with the publication of the Millennium Ecosystem Assessment by the United Nations, an appraisal, developed by scientists worldwide, of the role ecosystems play, directly or indirectly, in human well-being. Although buffered against environmental stress by culture and technology, humans are still utterly dependent on the flow of ecosystem services for our well-being, wrote the authors. This includes soil for food production, freshwater for drinking, wood for fuel, grass for animals, and open space for recreation. To make their point, the scientists grouped ecosystem benefits into four broad categories.

Provisioning services: food, fiber, fresh water, genetic resources, biochemical

Regulating services: climate, air and water purification, pollination, erosion control, disease and pest management

Supporting services: photo synthesis, soil formation, plant production, water and nutrient cycling (including carbon)

Cultural services: spiritual, religious, historical, aesthetic, recreational, educational, heritage, inspiration

The interaction and integration of these services in a specific ecosystem is critical. When they work in harmony with one another, human well-being rises; when they compete or damage one another, well-being declines—and not just for humans. Ecosystem degradation harms the well-being of multiple species. The erosion of wetlands, for example, can have cascading detrimental effects on a wide variety of plants and animals.

To no one's surprise, the UN assessment concluded that the demand for many ecosystem services around the globe is unsustainable. "If current trends in ecosystem services are projected unchanged to the middle of the twenty-first century," wrote the authors, "there is a high likelihood that widespread constraints on human well-being will result."[2]

"Constraints," I suspect, is a polite way of putting it.

Specifically, the authors say that the rapidly growing demand for provisioning services, such as water, food, and fiber, has been largely met at the expense of supporting, regulating, and cultural ecosystem services. Increased crop yields in industrialized nations, for example, have come at the expense of soil fertility, widespread erosion, and increased fossil fuel use. These costs have important feedback implications for ecosystem health and the services it provides. Also important, though less obvious, is the role rapid loss of culturally valued landscapes has played in social stress in many parts of the world. This is less apparent because the understanding of the linkages between ecological processes and social processes and their intangible benefits, such as spiritual and religious values, is weak.

In any case, it all adds up to a strong sense of urgency. Quickly reducing and reversing ecosystem service decline is necessary if we are to maintain the level of well-being to which humans are accustomed.

Blue carbon is a perfect example. The ecosystem services provided by resilient coastal and marine habitats run the gamut: storm surge protection, food security, water quality, recreational opportunities, poverty abatement, employment, tourism, shipping, fishing, cultural history, musical inspiration, spiritual values, sense of place, and on and on. Human well-being in southern Louisiana is inextricably linked to the services provided by its coastal ecology. When degraded—as they have been for decades—these services in turn degrade the quality of life for humans, as Hurricane Katrina exposed so tragically. When restored to health and functioning properly, the services provided by wetlands, swamps, and bayous can greatly improve not only our well-being but the well-being of animal and plant life too.

Did I forget to mention mitigating climate change as an ecosystem service? It just so happens that blue carbon stores large quantities of carbon, both in the plants and in the sediment immediately beneath them. According to scientists involved in the International Working Group on Coastal "Blue" Carbon, which met for the first time in 2011, total carbon deposits per square kilometer in coastal systems can represent up to five times the carbon stored in tropical forests—a result of their ability to sequester carbon at rates up to fifty times those of tropical forests. Occupying only 2 percent of seabed area, vegetated wetlands represent 50 percent of carbon transfer from oceans to sediments, the scientists said. In many cases these soils have been continuously building for five thousand years or more, and carbon stored in these sediments can remain sequestered for millennia. Salt marshes also have the added advantage of emitting negligible quantities of methane, they said.[3]

It's all about life, in other words.

Life is well-being: how we feel, how healthy we are, what we do, how we behave. When our well-being declines, life declines, especially if our attitudes sour into misery and anger. When our well-being improves, life improves. We humans, like all other species, have our own interests at heart, which means our well-being: food, shelter, water, family. To be sure, we're more than a collection of self-interests, thankfully, but if our well-being becomes stressed, then our other values often become subordinated to basic needs—such as survival. That's pretty straightforward, I realize, but what's always baffled me is why we don't take better care of ourselves. Why, if well-being is so important, are we so unwell in so many aspects of our lives? I understand that this is a complicated issue, but as I traveled around southern Louisiana I kept asking myself questions: Why did we put ourselves in this pickle? Why do we do things that damage our long-term self-interests? Don't we like life?

Take carbon. Not only is it essential to life, it is essential to our well-being—a critical element in abundance, coexistence, and resilience. But we keep doing things that damage the carbon cycle, which means we are damaging our well-being, and thus jeopardizing life itself. *Why?* It doesn't make any sense, not common sense anyway. Our behavior is crazy and bewildering. Still, at this point in my travels I knew some things. Thanks to science I had a pretty good picture of what carbon is, and because I'm human I have a good sense of what well-being means. What wasn't so clear, however, was the third part of the equation. I decided to dig. What, I wondered, is life?

Here's a quick answer: *life is anything alive.*

It is anything biological that can sustain itself over time, whether by respiration or reproduction, including growth, adaptation, and self-organization. Almost universally, life accomplishes these goals by employing cells—the smallest unit on the planet classified as a living thing. All cells have the same essential parts: interior and exterior membranes that regulate molecular traffic into and out of the cell; proteins that catalyze chemical reactions; a "library" of information in the form of DNA, which the proteins continually consult; and RNA, errand-runners who provide blueprints for the formation of new proteins. A cell is a complete package—it has everything it needs to grow and reproduce, provided it has access to minerals, water, and energy (carbon) in its immediate environment. The chemical process that enables a cell to transform these elements and energy into action is called *metabolism*. Its presence, along with replication and evolutionary change, is the foundation of life on Earth.

Recently, however, scientists have developed the ability to transfer DNA from one cell to another, changing its genetic makeup and creating an organism that didn't exist before. This new cell functions exactly like every other cell found in nature according to the definitions of life, but is it life? Yes, say scientists. Not so fast, say philosophers.

Is a computer alive? After all, a computer is a type of cell. It has a membrane through which energy and bits of data flow; it has DNA-like coded instructions in the form of programs and files that are constantly being changed and updated; its codes and files can be copied and shared with other computers; it has RNA-like wiring that carries electrical messages; it has a kind of metabolism, consuming electricity from its environment, generating paper printouts, and creating heat as a waste product. And there's lots of carbon in a computer—silicon too. By one definition, it's a carbon-and-silicon-based life-form!

Of course, a computer is not alive. For one thing, it can't reproduce itself (fortunately for computer manufacturers). Cells make copies of themselves, which is how an organism grows. Computers cannot do this—not yet anyway. Robots might be a different matter. Science fiction is littered with dark fantasies about self-reproducing robots run amok, usually at the violent expense of humanity. Is this a possibility? Here's a list of what biologists consider the basic ingredients of life: living things take in energy, they get rid of waste, they grow and develop, they respond to their environment, they reproduce and pass their traits on to their offspring, they evolve in response to their environment. Could be a robot! But are robots alive? It's another philosophical conundrum, although perhaps a moot one if a robot was coming at you with a laser gun in its (inorganic) hand!

What, by the way, is death? The opposite of life? The cessation of being alive? Scientists are messing around with this definition as well in their work, irritating the philosophers once again.

Here's what I think: *life is a force.*

About four billion years ago, against all conceivable odds, life came into being on Earth where no life existed previously. Chemistry yielded biology, and once life gained a perch it resiliently clung on, enduring billions of years of environmental stress. Seas boiled and froze; land flooded, rose, sank, and rose again. Oxygen levels—essential to life—were dangerously low for much of Earth's early history, an issue that was only resolved in favor of existence by the miracle of photosynthesis, an invention of life. Life perpetuating life. Five times over ensuing eons, life suffered massive extinction events, including one 270 million years ago that wiped out 80 to 90 percent of all living things on the planet. Still, life endured. It bounced back. It flourished again. Life is a force that won't be denied. Undaunted, life pushes on, urged forward by evolution and carbon, overcoming whatever physical challenge or toxic chemistry chance or circumstance can throw at it. It survives and thrives because it has one overriding purpose: to keep living. To keep going; to adapt, change, respond, be resilient—and not ever stop doing so.

Look at the history of life on Earth. In the beginning, there was none—zilch, nada. Today, there are between three million and thirty million different living species, biologists tell us, out of a grand total of as many as *four billion* species that have ever existed on Earth. Plants and invertebrates comprise something like 97 percent of all species alive today, including thirty thousand plant species, seventy-five thousand species of arachnids, one hundred thousand different species of mollusks, and anywhere from one million to twenty-nine million separate species of insects. Only 3 percent of all species alive are vertebrates, including eight thousand different reptiles, nine thousand species of birds, twenty-three thousand fish species, and nearly five thousand mammal species—of which humans are just a single one! And that's just what's still living. Biologists calculate that 99 percent of all species that have ever existed on Earth are now extinct, which means that death is a force that won't be denied either.

For much of Earth's early history, life was prokaryotic bacteria, which biologist Richard Dawkins has called "nature's superbly versatile chemists."[4] But things were bubbling and brewing, and roughly two billion years ago life begat the nucleated cell, the source of every species that followed—the foundation of what biologists call the tree of life. This tree includes us, by the way, out on the tippy, leafy end of an obscure limb, high, high up.

The first big branch to pull away from the trunk of the tree of life was plants, thankfully. Plants are the foundation of all food chains, which means almost all biomass on Earth comes ultimately from the sun via photosynthesis (and carbon). Don't forget the oxygen plants make. No plants, no life.

Next to branch away were fungi, in a split that took place approximately 1.4 billion years ago. Many people probably don't realize that fungi are more closely related to animals than plants (that is, they have a parasitical relationship to sun-loving plants, just like us animals). Next to split away were the protozoans, followed shortly by sponges (which are animals), sea anemones, corals, jellyfish, freshwater hydras, and flatworms, which meant the planet's volatile oceans were full of life more than one billion years ago.

Evolution, however, was just warming up.

About 590 million years ago a massive split in the tree of life occurred when the protostomes ("mouth first") separated from the deuterostomes ("mouth second"). The former included what became modern-day shrimp, ants, leeches, spiders, centipedes, insects, mollusks, brachiopods, and lots and lots of worms. The latter group included everything else, including us. Our common ancestor was very likely a dull worm.

Fifty million years later, conditions became optimal for a massive bloom of life. Called the "Cambrian explosion" by scientists, it was characterized by the sudden appearance of complex animals with mineralized skeletons, as found in the fossil record of the Cambrian Period. Research has revealed it to be a biological eruption of extraordinary diversity and quantity ("radiation" is the scientific term). All major animal shapes and most of the major animal groups that we know today appeared during this unprecedented event. To many researchers, it was the most important evolutionary transformation in the history of life on Earth, which is why it is sometimes called the "biological Big Bang." This event didn't take place overnight—ten million years is more accurate—but in geological terms it was still a blink of the eye. And it was never repeated again on this scale.

Along with all this biological diversity came a radical new ecological development called *predation*. The fossil record clearly shows that some creatures were hunters and some were prey—a development that had profound evolutionary consequences for life from this point forward. Ecosystems became much more complex as a result, and many animals moved (or were chased) into a variety of new marine habitats. Soon, Cambrian seas teemed with animal life of various sizes, shapes, and ecologies; some lived on the sea floor, while others swam around in the water. By the end of the period, a few animals had also made revolutionary first forays onto land (to get away from something hungry?), changing life on Earth profoundly once again.

Why did life explode like that? Some scientists point to a rise in oxygen levels that started around seven hundred million years ago, which might have provided fuel for an evolutionary explosion. Others believe that a biological extinction event just prior to the start of the Cambrian opened up ecological niches for new creatures, the way that mammals filled the big niche left by the sudden extinction of dinosaurs sixty-five million years ago. Others point to the "stitching together" of the supercontinent Gondwana at this time (which comprised today's South America, Africa, Antarctica, and Australia). Then there was the so-called carbon anomaly at the Precambrian-Cambrian boundary, in which the normal ratio of carbon isotopes in the carbon cycle was dramatically upset by some event. Some researchers say the animals themselves were responsible. One of the evolutionary consequences of predator-prey behavior, for example, might have been the development of shells and bony skeletons for protection. Maybe creatures were forced into "marginal" ecological niches where they had to adapt to survive, creating new body types where none had existed previously.

Whatever the reason, life was off and running.

The next branch away from the line that led to *Homo sapiens* included starfish, sea urchins, sea cucumbers, sand dollars, sea squirts, lancelets (a wormlike fish), and lampreys and hagfish (eel-like creatures). Then, 450 million years ago, sharks, rays, and ray-finned fish took off, including the ancestors of salmon, perch, cod, hake, carp, minnows, herrings, pike, and flounder. Amphibians separated next, around 340 million years ago, during the oxygen- and vegetation-rich Carboniferous Period. These included frogs, toads, newts, and salamanders. This was another important moment in evolution. Our common ancestor—what Richard Dawkins calls a "concestor"—probably looked like a salamander, though we don't know for certain. What we do know is that this concestor was the parent of the huge herd of land-based vertebrates that followed, including all tetrapods ("four-legged") on Earth—including us, at least before we evolved to walk upright. Dawkins calls this important concestor our 175-millionth great-grandparent.

The Sauropsids branched next, 310 million years ago. This is a large group that includes dinosaurs, reptiles, birds, turtles, iguanas, and crocodiles (our concestor in this era looked like a lizard, according to Dawkins). About 130 million years later, after a major extinction event (the one that wiped out 80 to 90 percent of all species on the planet), the duck-billed platypus split off from our line, followed closely by the marsupials. At the 105-million-year mark, the Afrotheres split off, which include elephants, manatees, aardvarks, and elephant shrews, as well as mammoths, mastodons, and other now extinct species (our concestor now probably looked like a shrew).

Anteaters, armadillos, and sloths split at the 95-million-year mark, followed by the Laurasiatheres, a huge group that includes modern-day moles, hedgehogs, bats, camels, whales, pigs, deer, sheep, hippos, horses, tapirs, rhinos, cats, dogs, bears, wolves, weasels, hyenas, seals, and walruses, among many others. This split took place in the "hot house" of the Upper Cretaceous, as Dawkins puts it, approximately 85 million years ago. It's an incredibly diverse group of herbivores, carnivores, and insectivores, with half of the group looking over their shoulders as they gallop or swim away, while the other half gives chase in an attempt to catch and eat them.

At the 75-million-year mark, the "gnawers" split off. This "rodent and rabbit" group includes beavers, rats, pikas, mice, gophers, marmots, hares, and porcupines. It's one of the great success stories of mammaldom—more than 40 percent of all mammal species alive today belong to this group. Our concestor at this point was probably another shrewlike mammal and is our 15-millionth great-grandparent.

At this point, we've reached the top branches of the tree of life, way up and off to one side—where our ancestors are hanging on with proto-simian hands, feet, and tails.

After the asteroid-induced mass extinction at what scientists call the K-T Boundary 65 million years ago, signaling the end of the age of dinosaurs, mammals essentially take over. Lemurs and bush-babies split from our line at the 63-million-year mark, followed by tarsiers. New World monkeys split next, leaping to another branch in the tree 40 million years ago or so. Our concestor here—the first anthropoid—likely lived in Africa at the time. Somehow, its descendants made it across the Atlantic Ocean to the New World, possibly riding on driftwood or some other type of raft, scientists speculate—a portentous sign of things to come!

Old World monkeys branched from our line 25 million years ago, followed by gibbons (18 million years ago), orangutans (14 million), and gorillas (7 million). Last but certainly not least, the final split—chimpanzees, at roughly 6 million years before present. This concestor, our last, represents our 250,000-thousandth great-grandparent.

That's four billion species in four billion years, give or take. Now, that's *life*—and all kicked off by a bit of interstellar carbon.

We are all stardust. Including bayous and jazzland bands.

Which brings me back to New Orleans.

The day after my tour with Mack and Day, I came across another answer to my question "what is life?" that I hadn't considered before—one that lifted my spirits considerably. It happened during another long ramble through the city, though this time I steered clear of as many street festivities, alcoholic hand grenades, and plastic beads as I could. I wasn't in a partying mood, to be honest, even from the sidelines. However, I didn't want to see anything that reminded me of being "6 feet under" either, so I avoided the river, the newspaper, and the Internet. I was tired, frankly, of feeling anxious and bewildered, weary of prognostications, reports, facts, calculations, analogues, and anything else that made me feel like I was falling out of an airplane without a parachute.

I aimed for the National World War II Museum instead.

I headed there for two reasons. First, I wanted to step back in time for a moment to an era that seemed, from a distance, simpler and more hopeful than today, a time when our nation had more purpose and direction than it does now, a time when the pickle didn't seem so large and daunting. I suppose much of the world back then considered Adolph Hitler to be a rather imposing problem, eventually requiring a catastrophic global war to solve, but he had the advantage of being a Bad Guy, which focused our attention and rallied our courage. Who is the Bad Guy today? Whom do we defeat? I've read countless calls for a "World War II–style" effort to combat the myriad challenges we face, but what does that mean exactly? I hadn't a clue, really, so I thought spending an hour or two soaking up the ambience and achievements of that bygone era would be educational and inspirational.

The second reason was more selfish. I wanted to know more about the world that nurtured my parents, members of the much-touted "greatest generation." Both of my parents died before I turned thirty—my father only three days prior. Over the course of my college years and after, we had never found much time to talk about their lives, where they grew up, what they did, where they went, or what they thought about their generation. As a typical twentysomething at the time, I was much more interested in *my* life, what *I* was doing, what *I* felt, and how much fun *I* could have. Roots? Memories? Photo albums? Family stories? No thanks, maybe later.

There wasn't any later, as it turned out.

Twenty years on, I'm looking for ways to make up lost ground. There was a story to tell about my parents and their times, I was certain—a story about two lives and a generation that spanned an extraordinary period of American history, from the Great Depression to the Apollo moon shot to, well, Homer Simpson. It was a story about hope and progress and the days before big pickles like climate change and rising sea levels. It was a story without a lot of detail, however. How did we get here? Who are we? Where are we going? I knew some of the answers, but much remained a mystery, including the period embracing World War II, an event that put an end to an earlier era and inaugurated a new one. I hoped to learn more at the museum.

I wanted to see that there was more purpose to our lives than getting out of pickles. Life had to be more than facts and numbers. Facts were good, but they only told part of the story. Life is music too, after all, and party beads, I suppose. It's more than just living, more than the sum of carbon, evolution, and our environment. It's more than resilience, more than survival and bouncing back. It's a force that wants to move forward, go places, be things; a force that can't be explained by chemistry or biology; a force that can't be stopped but can be shaped to do great things—if we do the shaping carefully.

What is life? Unexpectedly, I saw an answer on lampposts along a street as I rambled, where small, colorful banners waved. When I stopped to peer at one I saw that they announced an exhibit of paintings in town with a simple phrase that instantly lifted my spirits.

It said simply: LIFE IS ART.

That's exactly right, I thought. In our rush to engineer the world—and then to try to unengineer it when things go badly—we forget that at heart we are intuitive creatures, acting mostly on emotions. A scientist I know once said there is no "scientific method" to raising a child. He meant that while there are scientific aspects to child-raising, including visits to the doctor, in the end it's mostly intuition and experience. It's an art, in other words, not a science. Life is an art too, and if lived well it can bring great joy. Science can help (by calculating how many "hand grenades" are too much for our livers to handle, for instance), but in the end it is our emotions that matter most—how we respond to a good joke, a pretty song, or the presence of a wild animal. Of course, our emotions are just as responsible for the various pickles as our science, but facts alone won't show us the way home. They might even mislead us. We need art too.

In Carbon Country, we need both.

Leave It to Beavers

SHERRI TIPPIE, CREEKS ALL OVER COLORADO

Sometimes it takes a hairdresser to explain things.

In 2003, I invited Sherri Tippie, a self-described hairdresser from Aurora, Colorado, a suburb of Denver, to speak at a Quivira Coalition conference, where she urged the many ranchers in the audience to give beavers a chance. She knew the odds. For decades, the typical response of landowners to the presence of a beaver pond and gnawed trees on their property involved a gun, a trap, or a stick of dynamite. Beavers were considered "varmints" by many rural residents, who accused them of drowning productive fields, drying up irrigation ditches, plugging road culverts, and ruining valuable timber. And what do you do with varmints? You get rid of them.

"Don't," Tippie cheerfully told the crowd. "Beavers are your friends."

In a twenty-five year-career of rescuing and relocating the critters, Tippie has campaigned relentlessly to explain the positive things that beavers can do for landowners, which include reducing the risk of flooding, raising water tables, creating wetlands, and improving water quality and quantity—all good stuff in dry country like much of the American West. It's especially good if the country is suffering from a drought, as it is these days. Leaky beaver dams store water that would otherwise be lost—a message that ranchers and others are beginning to hear loud and clear as the West braces for a hotter and drier future.

Of all the good things beavers do, however, the least appreciated may be their role as *carbon engineers*. By one estimate, as much as 1 meter of sediment per year is caught behind beaver dams, and some sites can be occupied for as long as fifty years. Many dams are large as well, often stretching 1,500 feet. In 2010, researchers in northern

Alberta, Canada, discovered the world's biggest beaver dam, which at nearly 2,800 feet is twice the length of Hoover Dam! Beaver dams also create wetlands around their edges, which are well known for their carbon-storing capacity. For example, in the Upper Mississippi–Missouri River basin, researchers say, there were once over 50 million acres of beaver ponds. Although the total today is down to roughly 500,000 acres, that's still a lot of carbon sequestration going on—with the potential for much more.

Carbon sequestration wasn't something Tippie covered back in 2003, but it might very well be part of her pitch today. Growing up on the Front Range, she developed a love for wildlife at a young age. In 1981, in the midst of her salon career, she decided to act on her feelings by getting involved in efforts to stop state and federal agencies from poisoning coyotes near where she lived. Then one afternoon in 1985, "while scrubbing the floor with the TV on," as she described it, she heard a story about how Aurora had hired a trapper to kill beavers who were taking down trees on a city golf course. The newscaster said killing was necessary because there was no place to put them. The comment angered her into action.

She picked up the phone and in short order found a home for the beavers in Rocky Mountain National Park. When the state wildlife agency in charge of relocations complained about the time and cost involved in moving the beavers, she told them she'd move the animals herself. When the trapper complained that live-trapping beavers was impossible, Tippie told him she had seen one live-trapped on a National Geographic television show and that if he didn't want to try, she'd do it herself.

Which is exactly what happened.

"Don't underestimate the determination of a hairdresser with a cause," she told our appreciative audience.

Twenty-five years later, she's still just as determined, having live-trapped and relocated more than one thousand beavers, losing only two in the process (to a flash flood). She's still urging ranchers and other landowners to change their attitudes toward a species that once occupied nearly every watershed between northern Mexico and the Arctic Circle. She's still promoting the good things that beavers do and still pushing for coexistence in Colorado's creeks and streams. And her message over all these years has been the same: if possible, leave them be!

"When I started, I read everything I could get my hands on about beaver," she wrote in a recent article. "What I learned was that we really shouldn't be relocating them at all. When beaver are removed from good habitat they're simply leaving a void that other beaver will fill. The answer is to learn to coexist with them, because they are such an important species."[1]

Biologists have long considered beavers, in fact, to be a *keystone* species, estimating that 85 percent of all wildlife in the West at some point in their lives rely on the ponds and wetland habitat that beavers create. For example, beaver ponds are important nurseries for fish, including many rare and endangered species. And it's not just wildlife that benefit from our industrious friends. According to the Environmental Protection Agency, beaver ponds allow wetland microorganisms to detoxify pesticides and other pollutants, producing cleaner drinking water for people and reducing the cost of treatments downstream.

Conversely, when beavers are killed or repeatedly trapped for removal and the dams fall apart, a cascading series of unhappy changes occur, including decreased riparian stability, higher and more frequent flooding, reduced wetland acreage, degraded habitat for wildlife, diminished water quality, and less resilience to the effects of drought. That doesn't even count the large amounts of carbon that are transferred back into the atmosphere when wetlands and ponds dry up and their plant life oxidizes. Imagine the carbon consequences of having lost nearly 50 million acres of beaver ponds in the Upper Mississippi–Missouri River basin!

For nature, beaver ponds are an oasis of life. Not only do they provide drinking water for wildlife, but their still waters also harbor a wide variety of aquatic species. Their edges can be especially rich in plant life, including brightly colored wildflowers. Wetlands created by beaver dams are among the most biodiverse ecosystems in the nation, providing essential habitats for plants and animals that would otherwise struggle to survive. As oases, they also provide aesthetic and spiritual values to people, especially if it means getting a chance to spot an elusive beaver swimming to its den.

Fortunately, as I can attest, many of these arguments have begun to have an impact on ranchers and other landowners who now look on beavers as allies, not as enemies. And wildlife professionals credit Tippie, who is sometimes called the "Dian Fossey of Beavers," with much of this change in attitude.

"The amazing thing is, when people find out even a little bit about beaver, they often decide to coexist with them," Tippie wrote. "I believe it's my job to enlighten those people. Besides, it's fun!"

Judging from the various photographs I've seen of Tippie hugging a beaver, she's having a blast.

The beaver is the largest rodent in North America, weighing between 40 and 50 pounds, with a scaly, paddle-shaped tail and four buckteeth—two on top, two on the bottom. These incisors never stop growing, which means beavers need to keep them filed down by gnawing on trees and other woody objects. Beavers have webbed feet, dexterous hands, and transparent lids that cover their eyes when they swim. They also have a slick coat of fur and guard hair that enables them to live in a wide variety of ecosystems—a quality, unfortunately, that also made them valuable for their high-quality pelts, which were used, among other things, to make hats for fashionable Europeans.

We trapped *a lot* of beavers for their pelts early in our nation's history.

Before the arrival of Christopher Columbus, it's estimated that three hundred to four hundred million beavers existed in North America, or roughly ten to fifty beavers per mile of stream. Today, only six to twelve million beavers remain in their original habitat. The decrease wasn't only because of the demand for pelts and hats. In the 1820s, the British-owned Hudson's Bay Company sent trappers fanning out across the Pacific Northwest with orders to kill every beaver family they could in order to discourage American territorial ambitions. No beaver, went the company's logic, meant no "inducement to come hither," as one official put it. The result was the near extermination of beavers from an area the size of France.

The entire beaver trapping/killing episode was a tragedy of epic proportions, we're now realizing, and not just for its atmospheric consequences. Scientists directly link the removal of so many beavers across the American West to the widespread degradation of watersheds that we see today, which is why some consider their near annihilation to be the region's greatest environmental disaster.

Fortunately, it is a mistake that we are correcting, thanks to people like Tippie.

Which brings up another reason to put these wetland engineers back to work: adaptation.

From prehistoric times to the present, human societies have successfully adapted to the challenges of a changing region, including periods of extended drought, limitations created by scarce resources, and shifting cultural and economic pressures. However, the American West is entering an era of unprecedented change brought on by new climate realities—including greater variability in weather extremes—that will test our capacity for adaptation and challenge the resilience of the region's native flora and fauna. Beavers can help.

Beavers and their dams increase the ecological resilience of the land in the face of unanticipated changes. Here's a list of resilient attributes, borrowed from the Seventh Generation Institute, a nonprofit based in New Mexico that works to restore beavers to their rightful role on the land. A beaver dam:

- slows snowmelt runoff, which extends summertime stream flows and restores perennial flows to some streams;
- slows flood events, which could otherwise incise stream channels;
- contributes to the establishment of deep-rooted sedges, rushes, and native hydric grasses, which buffer banks against erosion during high flows and provide shade to creeks and streams, reducing water temperature;
- elevates the water table, which can subirrigate nearby land (including farmland);
- increases the amount of open canopy in forested areas;

- creates conditions favorable to wildlife that depend upon ponds, pond edges, dead trees, or other habitats in streams not modified by beaver;
- increases the mass of insects emerging from the water surface;
- creates favorable conditions for the growth of bank-stabilizing trees and shrubs, including willow and alder;
- greatly increases the amount of organic carbon, nitrogen, and other nutrients in the stream channel;
- ameliorates stream acidity; and
- increases resistance of the ecosystem to perturbation.[2]

As we enter a period of longer droughts, bigger floods, and rising demands for increased water quality and quantity by people and cities, competition among water users will only increase. Here's one simple answer: get beavers back to work.

To top it off, they do their carbon engineering for free!

What about the nuisances that beavers create, however, especially in road culverts? A beaver dam beneath a road can cause serious problems, as any landowner can tell you. For Tippie, the answer is easy: install a "Beaver Deceiver." It's a carefully constructed fence, narrow at the downstream end and wide enough at the other to discourage beavers from building a dam. Beavers are stimulated by the sound and feel of running water, which inspires them to start gnawing on nearby trees. The farther away a beaver can be kept from these stimuli by the fence, the more likely it is to be "deceived" into leaving the culvert alone.

The Beaver Deceiver was invented by Vermont resident Skip Lisle, who perfected the design as he beaver-proofed roads and other crossings on 120,000 acres of Penobscot Indian Nation lands in Maine. His deceivers are made with cedar posts, heavy-gauge wire, and a floor of fencing material, which blocks burrowing by beavers (who are very good diggers, by the way). The deceivers have proven to be very effective and are now in use across the nation.

Tippie is their biggest fan. "Skip's flow devices not only work, they take the needs of the beaver into consideration," she has said. "This makes them the best!"

She also considers them to be a work of art.

"I've seen flow devices that are ugly. They literally look like someone dumped construction debris into the water near a culvert or a beaver dam," she said. "Skip's flow devices are always well constructed, and this may sound strange, but they are beautiful."

So are beaver dams and the good work beavers do.

5

AFFLUENCE

The sound of falling water told me I had arrived at the right spot.

I had been walking up an old logging road in a lovely, narrow valley called Grassy Creek, admiring the late summer wildflowers and the bright blue sky, which at 9,000 feet above sea level seemed close enough to touch. A month of steady rain had turned everything green, causing an explosion of flowers and a big sigh of relief. It had been a year of exceptional dryness in New Mexico, part of a decade-long, low-grade drought afflicting the region. By late June where I live, outside of Santa Fe, the land had shriveled up and many of the plants looked dead, raising anxiety levels. We weren't alone in our concern. Across the state, farmers were abandoning their crops, ranchers were selling their cattle, and cities were imposing water restrictions on residents as the drought began to bite. The dryness even made normally robust waterways high in the mountains look tired and puny. Fortunately, July followed with record-setting rains, which, while not breaking the drought by any means, raised our spirits along with the wildflowers.

It wasn't merely the presence of falling water in the creek that enticed me, however, but the quality of its sound. It made a tinkling noise, like a musical fountain or a merry creek in a picture-postcard meadow. That was the rub. Tinkling wasn't a sound you heard often in Grassy Creek or any of the neighboring creeks where I was, in the Comanche Creek watershed, high in the Sangre de Cristo Mountains of northern New Mexico. If you heard water falling up here, it generally sounded hollow, dull and thick, as if it were being poured into a shallow drum—which it was, in a sense. Thanks to a century of hard use and damage by mining, logging, overgrazing, and an extensive network of poorly built roads, every creek in the 24,000-acre Comanche watershed has suffered a variety of eroding wounds, including numerous "headcuts"—small waterfalls in a creek that migrate upstream over time, draining water tables, ruining meadows, and undoing wildlife habitat. They are usually a sign of unhealthy land, and they often give away their location by their insalubrious sound. Water running over a headcut sounds hollowed out and drumlike, as if the creek had a chronic case of tuberculosis.

Merry it is not.

Some years earlier, I had hiked up a road that paralleled the main stem of Comanche Creek to its headwaters, and as I hiked I heard a veritable chorus of hollow drumming in the valley below me as water poured over one headcut after another. I felt like I had entered the TB wing of the watershed. That's how I knew I had arrived at the right spot in Grassy. The tinkling sound I heard told me this stretch of wet meadow had been cured; a headcut had been healed, the infection cut away.

It wasn't a random cure. Wet meadows are critically important in dry country like New Mexico, especially to wildlife, both aquatic and terrestrial, which makes them a priority for restoration projects. They are also great places to store carbon. In arid environments, such as much of the interior West, the carbon cycle operates more slowly than it does in wetter climes—for obvious reasons. Little rain means fewer plants, which means less carbon cycling—and a lot less when a drought is on. Thanks to the presence of life-giving water, carbon cycling happens faster in wet meadows and riparian corridors along creeks and rivers (as well as beaver ponds), which means you get multiple benefits when you cure these areas of what ails them, only one of which is carbon sequestration.

Wet meadows happen when a creek reaches a level spot in a valley, allowing its water to slow down and spread out. When healthy, these meadows tend to be spongy or soggy, which makes them highly productive, even in a drought. It also makes them a target for hungry herbivores, especially the domestic bovine variety, which, if left unmanaged, like to hang out in these meadows like idling teenagers, eating voraciously. Trails made by cattle along creeks as they travel to and from a wet meadow often capture the creek's water, which causes downcutting, which causes headcuts, which eventually drains the meadow. One solution is to remove the cattle or prohibit them from grazing near the water. Another might be to encourage beavers to move in (though wildlife are notorious for not following our instructions very well). However, once a creek catches TB in the form of a headcut, it usually requires professional medical attention to treat it properly.

Which is what we were doing that day.

Aiming for the tinkling sound, I left the logging road and crossed the small meadow. Reaching the edge of the creek, I saw the cure: four spruce logs carefully set in a Z pattern in the channel, which caused the water to slalom back and forth as it made its way down the 5-foot run. At the bottom, the water tipped over a final log to splash merrily onto two rows of logs set flush with the flow of the creek. Anchoring the ends of the logs along the creek banks were rocks of various sizes, each judiciously placed. I also saw green tufts of sod, cut from the meadow, arranged among the rocks in order to secure them the old-fashioned way—with green, growing things. With life. The net result of the logs, rocks, and tufts was this: the headcut had been replaced with the sloping log-and-rock structure whose zigzag design and soft landing dissipated the water's erosive energy, terminating the small waterfall's inexorable march upstream. In other words, the small meadow had been saved. This stretch of Grassy was healing, and already it was singing its merry song once again.

This was good, but it wasn't why I had hiked up the logging road.

The Quivira Coalition has been doing restoration work in the Comanche Creek drainage for a dozen years now in collaboration with the Carson National Forest and a host of conservation-minded organizations and agencies. Thanks to the hard work of a boatload of workshop volunteers, nearly two hundred in-stream structures have been installed, so there wasn't anything particularly "newsy" about the work at my feet, though I hadn't seen logs used in this way before. In fact, two crews of volunteers were working a short distance downstream from where I stood, healing two more head-cuts. They were here as part of an annual August workshop that the Quivira Coalition organizes in the watershed, and before we hiked up Grassy the leaders hosted a quick tour of previous work. I went along, and what I saw was an impressive amount of healing. One of the participants, a retired geologist who had been leading students on educational field trips to Comanche for a long time, said to me, "Courtney, twenty years ago this valley was all rocks and dirt. Now look at it! It's incredible."

He meant the grass and the wildflowers, but he also meant the power of life at work.

There were two principal goals to our restoration work in Comanche over the years. First was to field-test on a watershed scale the innovative ideas of Bill Zeedyk, a pioneering riparian restoration specialist. Over the years, Bill has developed an effective set of low-cost techniques that stop downcutting in creeks, often by "inducing" an incised stream to return to a stable channel through the power of small flood events. Bill calls it *induced meandering*. When a creek loses its riparian vegetation, it tends to straighten out and cut downward because the speed of water is now greater, causing the scouring power of sediment to increase. Over time, this downcutting results in the creek becoming entrenched below its original floodplain. Eventually, a creek will create a new floodplain at this lower level by re-meandering itself, but that's a process that takes decades. Bill's idea is to goose the process along with his structures.[1]

Our second goal was to assist the Rio Grande cutthroat trout, New Mexico's official state fish, which is in trouble. Reduced to something like 10 percent of its former range by development, predation by nonnative fish, degraded habitat, and water pollution, Rio Grande "cuts" are struggling to hang on. Toss in reduced stream flows as a consequence of the drought and projected rising water temperatures due to global warming (bad for native trout) and you have a recipe for some serious consternation among state biologists, federal land managers, fly-fishing aficionados, and everyone else who cares about wildlife. Fortunately, the two goals merged nicely. Bill's bioengineering approach to riparian restoration has helped heal Comanche Creek, which in turn has helped the trout (as I have described in my previous book, *Revolution on the Range*).[2]

What we didn't think about was carbon—at least in the climate context. Although I'm certain our restoration work is doing a lot to rebuild carbon stocks in the soils of formerly degraded wet meadows and eroded creek banks, I couldn't say precisely how much—because we didn't take any baseline measurements! There have been some studies recently that suggest that the carbon storage potential of stream banks is very good, especially if a healthy riparian plant community can become well established, or if beavers move in. We just didn't do any carbon accounting on Comanche Creek, which I regret. Sequestration wasn't on our radar screen back then.

However, I wasn't inspecting the log-and-rock structure on Grassy Creek for its carbon-sequestering potential, either. In fact, I wasn't there for science or medicine, restoration or climate mitigation. I wasn't so much interested in the structure's function as its form. I was there, frankly, to see if it was *pretty*.

Why? One of my heroes is the conservationist Aldo Leopold, widely honored for his pioneering work in many fields of endeavor, including game management, wildlife biology, wilderness protection, watershed restoration, and environmental education. He even put in a good word for sustainable agriculture. While he is best known for his articulation of a land ethic, which was essentially a plea for harmony between land and people, as well as the concept of land health, which encompassed the regenerative processes that perpetuate life, there is another aspect of his deep thinking that has been largely overlooked. It took root back in 1912, when he became supervisor of New Mexico's Carson National Forest, which at the time didn't include the Comanche Creek watershed. Leopold noticed that when land became "sick"—that is, when its basic parts fell into disorder or broke down, including high rates of soil erosion—it also became *ugly*. Health included beauty, he realized, which meant that aesthetics needed to be an important component of conservation and agriculture.

The best way to explain it is with Leopold's own words:

> There is only one soil, one flora, one fauna, and one people, and hence only one conservation problem . . . economic and esthetic land uses can and must be integrated, usually on the same acre. ("Land Pathology," 1935)[3]

> Bread and beauty grow best together. Their harmonious integration can make farming not only a business but an art; the land not only a food-factory but an instrument for self-expression, on which each can play music of his own choosing. ("The Conservation Ethic," 1933)[4]

> Who is the land? We are, but no less the meanest flower that blows. . . . What are the sciences? Only categories for thinking. . . . What is art? Only the drama of the land's workings. ("The Role of Wildlife in a Liberal Education," 1942)[5]

> One cannot divorce esthetics from utility, quality from quantity, present from future, either in deciding what is done to or for soil, or in educating the persons delegated to do it. All land-uses and land-users are interdependent, and the forces which connect them follow channels still largely unknown. ("Deer and Forestry in Germany," 1935)[6]

> I suspect there are two categories of judgment which cannot be delegated to experts, which every man must judge for himself, and on which the intuitive conclusion of the non-expert is perhaps as likely to be correct as that of the professional. One of these is what is right. The other is what is beautiful. (letter to the *Journal of Forestry*, 1935)[7]

Art, harmony, beauty, aesthetics—as perceived by us—are all signs of health in nature and ourselves. Conversely, ugliness is a sign of ill health, whether it is a suppurating hard-rock mine, an eroding stream channel, or harsh words at a public meeting. I had experienced all three, and much more, during my time as an environmental activist with the Sierra Club. I learned that while stopping bad things was necessary, negative energy rarely led to positive results. Ugliness mostly begat more ugliness. This was the main reason I cofounded the Quivira Coalition—to inject positive, collaborative, healing energy into the very ugly brawl between environmentalists and ranchers. It was also one of the reasons I took a shine to Bill Zeedyk's ideas about creek restoration when I came across them. Bill's methods, much like the ranching practices we advocate at Quivira, harmonized with the land and its ecological processes. Not coincidentally, his structures were also attractive to look at. Made of rocks and wooden posts and designed to direct water naturalistically, they had a sculptural feel that verged on the artistic. It was work that integrated form and function—just as Leopold had hoped.

I turned and headed upstream.

The creek steepened, and after a short hike I came to another log-and-rock structure. Like its downstream kin, I heard it before I saw it, but instead of a merry tinkling sound, this one sounded like a small cascade. That's because the structure—called a "step-down" by its inventor and installer, Craig Sponholtz—was *huge*. It stretched 30 feet along the creek, replacing what must have been a giant headcut. Water didn't so much slalom as sluice between the dozen or so logs as it made its way to the double row of logs at the bottom. Additionally, a small channel of water entered from the left, sauntering down a deliberate course before plunging softly into a rock-lined bowl, where it paused for a moment before spilling into the creek below. Large rocks lined the creek's banks on either side of the step-down, and beyond its upper end I could see a long meadow extending into the distance. It looked dry, and as I stood there I thought I heard echoes of its coughing in my ears.

I turned my attention back to the step-down—a new design by Craig that was making its debut right here in Grassy Creek. We've worked with Craig for six years now, and I knew he was both inventive and artistic. But the step-down knocked me out. Not only was it impressively constructed, it was lovely to look at. Craig had arranged the zigzagged spruce logs in the creek to make it look like the trees had simply toppled over from the nearby forest instead of having been carefully placed by a machine (an excavator in this case). The locally sourced rocks had also been fitted around the logs in a way that was pleasing to the eye—and yet I knew each one had an important role to play in healing the creek. I also knew there was a substantial amount of heavy-duty filter fabric underneath the rocks and logs to hold the soil in place, even though not a stitch of it was visible on the surface. This was by design as well. Add in the tufts of sod inserted between the rocks and logs, the pretty rock-lined bowl at my feet, and the burbling sound of cascading water, and you had the recipe for a Zen-like work of art—a work of art at 9,000 feet, in a wildflower-strewn valley, nestled between a rock outcropping and a patch of spruce forest, under a canopy of blue sky.

What a gallery for a grand opening!

Of course, the step-down structure had a job to do, first and foremost. Its assignment was to save the wet meadow by easing water down a steep stretch of creek (formerly the headcut) without incurring any additional erosion, especially in the event of a big rainstorm. Accomplishing this goal requires knowledge of soils, hydrology, geomorphology, mechanical engineering, and math on the part of the designer, as well a great deal of field experience (and a soft touch with an excavator), or the structure will fail in its duty. But this is where Leopold came in. What Craig had done here—as he has done in many other projects—was take something totally functional and entirely human-constructed and make it look like a natural feature on the land, in this case an attractive log-filled cascade of merry water. It was a wholly practical restoration structure *and* a piece of sculpture. The dictionary defines *sculpt* as "to carve," which is exactly what Craig had done—carved the land into health and beauty.

That's why I hiked up Grassy Creek—to explore my hunch that the principles that made a practice restorative and regenerative were the same ones that made it beautiful. It wasn't form *or* function, art *or* utility, it was both—as I had seen in the carbon ranching ideas of John Wick and Jeffrey Creque, the progressive cattle management practices of Tom Sidwell, Dorn Cox's vision of abundance, Colin Seis and Eric Harvey's belief in coexistence, Sarah Mack's plans for resilience in southern Louisiana, and all the other stops so far on my journey. There was science at work, of course, as well as intuition and experience, but there was beauty too—which is why it's so often called the "art" of ranching or farming or land restoration. There's something that binds them all together. Leopold had his finger on it when he wrote in *A Sand County Almanac,* "Land is not merely soil; it is a fountain of energy flowing through a circuit of soils, plants, and animals."[8] When this energy is positive, good things happen, as I've discovered.

And at the heart of this fountain is carbon.

Leopold is an inspiration to Craig as well. His favorite quote is this one: "A thing is right when it tends to preserve the integrity, stability, and beauty of the biotic community. It is wrong when it tends otherwise." That's exactly what the log-and-rock step-down was doing—restoring the integrity, stability, and beauty of the biotic community known as Grassy Creek.

I peered around for the artist, spotting him a short distance away, carrying a chainsaw in a case and leading two young volunteers across the meadow. Wearing sunglasses and a straw cowboy hat and sporting his familiar goatee, Craig was hard to miss. Tall and broad-shouldered, at forty he still had the upright bearing of the wildland firefighter he once was. No wonder he looked at ease carrying the chainsaw (he confirmed later that one of the reasons the step-down had been fun to create was because it allowed him to use his timber-cutting skills again). He gave me a wave, indicating he'd be right over. One of the volunteers carried a tripod and a monitoring device of some sort, and I suspected they were preparing to take baseline measurements of the restoration work. These projects require a ton of documentation, I knew, for both the funder and the landowner—in this case the state of New Mexico and the US Forest Service, respectively. Data also helps chart the progress of the healing process. Art, meet science.

A few minutes later, I met Craig at the head of the structure, where I had been admiring the rock work. I had noticed that some stones had moss on them and they had been carefully oriented so the moss would be plain to see. I asked him about it.

"If the moss didn't show," Craig said, "you'd notice. Your eye would catch it as you looked around."

I played devil's advocate for a moment. "Does that sort of thing really matter?" I asked. "Is it necessary to be so artistic? After all, whether the moss is upside or down will have no bearing on the function of the step-down. Why make it pretty?"

"Beauty forms a large part of our relationship to nature," Craig replied, "and we react negatively when it's absent, as with degraded landscapes. But I understand what you're saying. Beauty alone doesn't heal wounds."

He nodded toward the meadow.

"But if you're going to spend time and money trying to heal a meadow like this," he continued, "which is critically important to the ecosystem, then I think it's best to follow nature's blueprints, which involves an intricate web of interactions that life depends on. Beauty is part of that web, as are water, soil, and plants. You can't have one without the other."

We took a few steps into the meadow. Although he needed to get back to the volunteers, he wanted to show me how he had diverted part of the creek's flow into the side channel, where it eventually made its way into the rock-lined bowl at the bottom of the cascade. The goal was to reduce the erosive pressure of the water in the main channel and saturate the adjacent meadow. The rocks and logs in the step-down were secondary, he said, to understanding the way water moved across the meadow and reached this point.

"It's all about the flow," he said. "Trying to understand where the water comes from and where it wants to go. You have to know this before you begin your design, or else it won't work."

As Craig spoke, the word *affluence* popped into my head. In hydrology, it's a technical word for a tributary stream, such as Grassy Creek. In creek restoration, it means an abundant supply of water, flowing freely toward a point (its linguistic roots are Latin via Old French meaning *fluid*—I looked it up later). In other words, you *want* a creek to be affluent, especially if it's been degraded or damaged by historical use, as nearly all the creeks in the Comanche Creek drainage had been. Craig's work, and the goal of our workshops, aims to do exactly that—to restore affluence to the land, to make it rich again.

The art of restoration, Craig said, isn't simply about the structure, the materials used to make it, or its intended effect on the creek. It's also about what happens before you get started. It's about recognizing opportunities, which requires careful observation and a clear understanding of nature's regenerative principles. Opportunities can take the form of a boulder or bedrock outcropping, a clump of sturdy vegetation, a low bank, or almost any other feature that can be used to create a unique solution. Once recognized, he said, these opportunities help create a structure that harmonizes with long-term natural processes.

He calls these opportunities *sweet spots*.

"My goal is to recognize the small opportunities that make a big difference and to act on them," he said. "This is why watershed restoration is endlessly creative and endlessly rewarding."

A week later, I met Craig at another sweet spot, though this one was about as different from Grassy Creek as you could possibly imagine. Instead of wildflowers, a meadow, and running water, it featured a metal pipe under a road, a large field of rocks, and a bone-dry river. We were in the heart of Santa Fe this time. The pipe was part of the city's stormwater discharge system, the rocks were part of Craig's creative solution to an eyesore, and the waterway was the Santa Fe River, dry as a result of the drought. A year or so ago, Craig and a few colleagues had approached the city with their idea on how to solve the erosion problem associated with the discharge pipe, transforming a sore spot into a sweet one, and the city gave them the green light.

I parked on the street, which served a nearby housing subdivision, and walked the few feet to the project site, noticing that it was in a highly visible location adjacent to a popular hike/bike trail that paralleled the river. Craig said this was part of the project's appeal to him. It was a chance for him to create a pleasant place for families and others to relax—in a place that also fulfilled an important drainage service to the city.

We inspected the drainage pipe and the rock field first. The standard procedure for moving stormwater from a street to a river was to "riprap the hell out of it," as Craig put it, meaning build a large, U-shaped, cobble-filled wire structure that was neither pretty nor family-friendly. Stormwater is considered a nuisance to most city managers, and riprap takes care of the problem with brutal efficiency. Craig had a different attitude toward the water that emerged from the storm pipe: use it as a resource. The first thing he did was analyze the pipe's "watershed," which in this case meant the roughly 80 acres of streets and homes behind us that drained into the storm system.

By Craig's calculations, 80 acres meant a decent amount of water flowed out the pipe, including snowmelt, and although the water flowed intermittently, it was more than enough to potentially nurture a bevy of native plants and trees. In other words, instead of being a nuisance, the stormwater could become a type of "urban spring," as Craig described it, used to grow things. His goal was to develop this "spring" into something lovely for the city. Instead of riprap, Craig created a rock cascade below the drainage pipe that broadened out as it approached the riverbed. Part of its purpose was to move water away, but another purpose was to let water sink in—to water the numerous plants and trees Craig and his crew had sowed in the ground. Instead of straight and ugly, the cascade had an intentional tumble-down look to it, as if the rocks had been placed there by nature as part of a regular watercourse. It was pretty, but it also had work to do.

"It's called rock mulch," Craig said, "and its job is to slow runoff, increase soil moisture, protect seedlings, and retain soil particles. You'd be surprised how much soil is mixed in with stormwater runoff."

Ringing the rock cascade were several large, flat rocks—perfect for sitting on—and dotting the project site were young trees, mostly chosen by Craig for their ability to provide shade in just a few years. Less obvious than the trees were the subtle mechanisms by which Craig moved water to the river. These included a layer of bark and wood chips, all hidden from view. As a final touch, surrounding the stormwater pipe was a cement wall with a brightly colored mural of green plants set against a yellow background. It had been painted by a friend of Craig's with the goal of setting a festive mood. It was working, too—people were already using the project site as a place to rest and recreate.

When the project was completed, the city of Santa Fe held a public ceremony to dedicate what was essentially a minipark along the hike/bike trail. It was part of a larger celebration for a myriad of restoration projects that had been completed in the river channel. Various dignitaries attended, including the mayor, who praised the stormwater project as a new way to deal with an age-old problem. He noted that except for the presence of the storm pipe, you would never have known that the little park had a highly utilitarian function.

I asked Craig if this sweet spot was a role model. Would the city begin to rehab other stormwater sites into urban springs?

"I don't know," Craig replied. "Something like this isn't as cheap as cement and riprap, so who knows what they'll do in the long run. But I hope this will light a fire."

One place Craig definitely lit a fire was out in the dry, sparse country of the Ojo Encino Chapter of the Navajo Nation, west of Cuba, New Mexico. In an effort to rejuvenate their overgrazed and eroded land and create a brighter future for themselves, members of the chapter embarked in 2005 on a wide-ranging plan of ecological and economic restoration. As part of this plan, they asked Craig to teach a series of erosion control workshops for crews of young Navajos as part of a summer training program. When he was done, he stood back and watched as they took responsibility for site selection, treatment design, and construction of a variety of erosion control structures across chapter lands.

"It was really great," Craig said. "Not only was the rock work outstanding, but the designs laid out by the crews demonstrated their real understanding of the way water and sediment move across the landscape."

Another project involved abandoned corn fields.

Due to erosion and poor water management, many traditional corn fields in Navajo country had been abandoned. During an Ojo Encino–led workshop, I had the opportunity to see one myself. A deep gully, created by a poorly designed and maintained dirt road, had charged straight downhill to the corn field, sinking deeper into the land as it went, draining the water table and ruining the field for agricultural purposes—and ruining its affluence as well. Not only was everything dysfunctional, it looked ugly too. Similar gullies afflicted corn fields across the sagebrush-dominated landscape, threatening the ability of an already impoverished people to feed themselves. These fields were redeemable, however, said the workshop leaders, if the gullies could be fixed and the water cycle repaired—which is what the chapter subsequently set out to do with the help of the young restoration crews.

Both projects were examples of the chapter's effort to restore *hózho* among the people, wrote community leader Tammy Herrera later in an article for Quivira's journal. *Hózho*, she wrote, translates as "walking in beauty." It means to live in a manner that strives to create and maintain balance, harmony, and order. "This single word captures the essence of Navajo philosophy," she wrote. "It is similar to but much richer than the term 'conservation' as it implies a deep connection between land and people. One cannot be restored without the other."

Hózho is similar to Aldo Leopold's famous concept of a land ethic, she wrote. *Hózho* is all about restoring land health, reconnecting people to their land, healing wounds, feeding the community, maintaining traditions, reengaging youth, and building local capacity to make communities viable and resilient in the face of multiple challenges—including chronic poverty and now climate change. For Tammy and the six hundred members of the Ojo Encino Chapter, all of this actually means rediscovering *hózho*—because for many the idea of walking in beauty has eroded along with their corn fields. For example, federal policies have forced many Navajo into the commodity food system, which has resulted in skyrocketing rates of poor nutrition and diabetes.

"Over the past sixty years our people have become almost entirely dependent on the outside world for everything from food to clothing to fuel," she wrote. "As we have gradually moved away from a lifestyle in which we depended on the land for all of our basic needs, we have shown a significant decrease in our care for that land . . . as a result, the land has suffered. The relationship has been broken."

To restore *hózho*, Tammy and fellow community members have (1) written and implemented new grazing plans for their livestock; (2) embarked on an ambitious range improvement program (for instance, tackling the ubiquitous sagebrush); (3) implemented an innovative feral horse population control program involving a new contraceptive drug for mares; (4) tackled erosion challenges caused by poorly designed and maintained dirt roads on the reservation; (5) trained crews and young leaders in various land restoration techniques, including the workshops that Craig led; (6) hosted native food fairs; and (7) initiated an effort to revitalize old corn fields, among other efforts.

"There is much to be optimistic about!" Tammy wrote in the conclusion of her essay. "At meetings we talk about healthy food, healthy livestock, and healthy land. We complain less and work to solve our own challenges. When people come from the outside and talk about the need for grazing plans, we proudly show them our plans. We have youth who show up in the summer to haul rocks to erosion sites even though they know they aren't on the payroll. Each of these things is beauty. . . . We are creating a new way of life that combines the old with the new. There is no guide book. Each small step is a major victory and a step closer to what is right, to *hózho*, Walking in Beauty."[9]

A goal for all of us, I think.

So, how did Craig, a former Forest Service firefighter, became a restoration artist?

Born in Oklahoma and raised in Denver, where his father worked as an FBI agent, Craig developed a love of the outdoors at an early age. His interests included camping, hiking, and snowboarding—what he calls that "Colorado thing." But he also developed an interest in landscaping, which he did for pay after high school, and it was on these projects that the idea of ecological design began to tickle his imagination. After high school, he did a stint as a wilderness ranger on a national forest near Sante Fe, which rekindled his love of the outdoors. A move to Silver City, New Mexico, for college came next, followed by a job on a forest trail crew making illegal or badly designed campsites "disappear"—which got him thinking once more about ecological design. Next came a turn as a firefighter with the US Forest Service, followed by a promotion to an elite "hot shot" crew, which often tackles some of the toughest fires. He loved the work, as well as the "outstanding" pay, as he put it.

Meanwhile, he was trying to finish college at Western New Mexico University, describing himself as "not in a hurry to graduate." He took forestry classes mostly, knowing he was being groomed for a career in the Forest Service, but he didn't feel a rush to join the federal workforce. In fact, he began to suspect that something was missing from his life. To find out, he enrolled in art classes as well, reactivating an interest he had developed in high school, taking courses in printmaking, ceramics, photography, sculpture, and welding. He was particularly fascinated by the process of creating art—the *doing*—and not so much in grand visions. "I wasn't trying to make a statement about anything," he told me. "I mostly wanted to learn the technical steps. A lot of what I made was rather abstract."

Meanwhile, Craig had purchased a small piece of undeveloped land in the Burro Mountains, southwest of town. The property was worn out from hard use, Craig said, which is why it was cheap. He saw the land's poor condition as a challenge and knew he wanted to heal it and make it better. But how exactly? He started by building check dams in eroded gullies on the property (a big no-no, as he would learn later from Bill Zeedyk). He planted trees next, but that didn't go as planned. He felt like he was floundering. Then a friend gave him a book on permaculture, which is a design process based on ecological patterns found in nature. It was developed by two Australians, Bill Mollison and David Holmgren, in the 1970s, and Craig was so intrigued by what he read that he decided to head Down Under to take a permaculture class. It netted two results: first, abandonment of his "leave it alone" conservation philosophy that he had adopted as a wilderness ranger, which hadn't helped very much with his battered property in the Burros anyway; and second, a decision to find a career in something more creative than fighting fires.

Inspired by his permaculture experience, Craig decided to continue his education by enrolling in Prescott College, in central Arizona, where he custom-designed a master's degree in agroecology, a field that studies the ecological processes at work in sustainable agricultural systems. Under the guidance of professor Paul Sneed, whom Craig credits with urging him along intellectually, Craig studied water harvesting techniques used by prehistoric Southwestern tribes in their dryland farming practices, including mulch gardens and catchment structures. He was particularly interested in prehistoric methods of erosion control, thinking once again about his experience on his property in the Burros. His studies focused on how water moved across the land, how it could be stored and released for food production, how to avoid crossing ecological thresholds, and how to restore damaged land using the same principles. It was, he admitted, all about *function*.

Art would have to wait.

Finishing his degree, Craig quit his job with the Forest Service and moved to Santa Fe, hoping to find work in land restoration. At a workshop organized by a nonprofit called Earthworks on a property south of town (which I also attended), he met Bill Zeedyk for the first time and was immediately enthralled by Bill's methods of creek restoration. Craig quickly became an apprentice, soaking up every aspect of Bill's philosophy and experience. It wasn't just the design work that excited him, it was also the process of *doing* the restoration work that motivated him, bringing back his artistic training and love of form. At a subsequent training class led by stream specialist Dave Rosgen, who had been Bill's mentor, Craig was deeply impressed by the structures built by Rosgen's contractor, who did "beautiful work," as he put it. In turn, when Bill praised a structure that Craig had constructed in an eroded creek as "beautiful," it gave him a big boost of confidence. It also felt like he had turned a corner. With his apprenticeship coming to an end, he said, it was time to begin adding "flair" to his work.

"I believe Bill Zeedyk's greatest insight is that we must become partners with natural healing processes and that the art is to know which part of the process we must become," Craig said. "At times we can act as catalysts, jumpstarting the establishment of plants that will provide stability. Other times, it is only necessary to gently steer a process, as in favoring the natural erosion of one bank to build floodplain on another."

Fast-forward to today and Craig now has more than one hundred projects under his belt, many with the Quivira Coalition, and every one that I have seen has been amazing to behold. Especially lovely is the water-spreading, crescent-shaped, one-rock-tall structure called a *media luna* (half moon), which Craig has perfected into sculpture. Another specialty is an in-stream grade-control structure called a *cross vane*, which is composed of large rocks carefully arranged in the creek in order to slow down the water's momentum by creating a natural plunge pool. In Craig's hands—or more accurately his backhoe or excavator—the cross vane invariably becomes another Zen-like work of art.

There's a minimalism to Craig's work that is intentional, I've noticed. After taking care to read the landscape of the project site diligently, and thinking a lot about boundaries and edge effects, Craig creates a design that involves as few people and materials and as little dirt-moving as possible, while striving for a strong and long-lasting effect. The minimalism is partly about self-expression, he admits, but it is also about physical objectives—to heal the creek as simply and effectively as possible. It also makes sense economically, especially to the landowner or agency funding the work. Beauty is woven into the minimalism too, which accounts for the naturalistic feel of his structures.

"There's beauty in the process," Craig said, "but there must be beauty in the final product too."

Today, Craig calls what he does "regenerative earth art." Not only is his goal to heal damaged land and create affluence for anyone who lives in a watershed (all of us, in other words), he creates structures that become part of the ecological processes that they reignite. By serving as footholds for grass and riparian plants that take over, his structures eventually are absorbed into the land itself and disappear.

His regenerative design principles include the following:

- increase ecosystem services and productivity
- protect and expand natural moisture-storing areas
- stabilize erosion
- restore flow and infiltration
- cultivate regenerative plant communities
- harvest runoff
- transform problems into opportunities
- create solutions that use natural processes
- strive for beauty

As I said earlier, these principles fit very nicely with the goals of the carbon pioneers that I've met so far on my journey. Carbon sequestration, building soil, storing water, growing nutritious food, building resilience, and now restoring creeks to form and function—it's all one job. Toss in the concept of a sweet spot, where big things can happen in a small place in a hurry, and you have the outline of a plan of action. Better yet, all this work, taken together, can create a type of affluence that is critically required today—and not just the fluid variety (interestingly, the term didn't take on its connotation of material wealth until the sixteenth century). I mean *rich* in the old-fashioned sense of copious, bountiful, fruitful, prolific, vibrant, full-bodied, resonant, deep, vivid, elegant, fertile, and abundantly supplied.

Here's the kicker about restoration, however: we can all participate.

"The main misconception that people have about watershed restoration is that it's something that happens far away in parks and public lands and not something that can be part of everyday life,"

Craig said. "But everyone lives in a watershed, and I work hard to make the restoration of our home watersheds something that is built into the ways we live and work."

Recognizing the signs of trouble is the first step, he continued, whether it is poorly drained roads, eroding fields, unhealthy creeks, poor stormwater management, or undervegetated land. The second step is to learn what the proper remedies are for these ills so that we can make informed decisions about how to make the land, and our lives, richer and more abundant. In particular, the nexus between agriculture, water harvesting from buildings and roads, and stormwater management, Craig believes, is directly connected to human well-being.

"As agriculture expands its footprint on the earth to feed the billions here and billions on the way, our watersheds are suffering and we are leaving less and less space for nature to play out critical natural processes," he said. "The same is true in many urban areas, where runoff is shunted away to the nearest drainage, carrying garbage and pollutants. This truly diminishes our quality of life and makes our lives and our connections to the land less resilient."

The key is informed observations about the land. Too often, we jump at solutions before we have a clear understanding of the problem. This is what happened to Craig on his own little property in the Burros, where he struggled alone through eight years of trial and error in an attempt to solve a problem that didn't really exist. For example, what he thought was an arroyo (dry streambed) on his land was actually part of an alluvial fan. This was important because you can't treat an alluvial fan like a regular arroyo and expect it to remain stable. From years of experience, he has learned that if you try hard enough to solve a nonexistent problem, you can certainly go a long way toward *creating* one.

On the flip side, knowledge can bring heartbreak. As Aldo Leopold famously observed, "One of the penalties of an ecological education is that one lives alone in a world of wounds."

It's a world that Craig knows all too well.

"When I started down my own path, I observed so many wounds, some superficial and some apparently fatal. I only knew enough to recognize the wounds," he said. "My unquenchable desire to do better by the land eventually taught me that all those wounds are opportunities yet to be realized. They are opportunities to connect with the land, with complicated processes, with natural beauty, and ultimately with each other."

This brought up a question I've had on my mind for a while: how many people could recognize an ecological wound if they saw one? Most of us would recognize obvious ones, probably, but what about subtle ones? Could many people tell a natural arroyo from an eroding gully, or what either might mean for the health of their land? Could we tell if plant pedestaling was a sign of proper land function or a sign of erosion? If we recognized a headcut in a wet meadow, would we be able to deduce why it was there or where it originated? Could we tell if a channel was aggrading or degrading, or why we should care? Do all creeks meander? And if one does not, what is it telling us about the state of its health?

In other words, how is our *land literacy*?

This issue hit home for me years ago when I heard Dan Dagget, an environmental activist and early Quivira Coalition supporter, tell a story about a professor of environmental studies he knew who took a group of students for a walk one day in the woods near Flagstaff, Arizona. Stopping in a meadow, the professor pointed at the ground and asked the students, not so rhetorically, "Can anyone tell me if this land is healthy or not?" After a few moments of awkward silence, one student finally spoke up. "Tell us first if it's grazed by cows or not," he demanded. The implication was clear: If cows grazed there, the land had to be unhealthy. If cows did not graze there, then things were "natural" and therefore fine. Dan's point was that the actual condition of the land, visible as signs of health or ill health, had become secondary to the political positions of the observers. The point that stuck with me over the years, however, was this one: we've become mostly illiterate when it comes to reading a landscape.

There's a good chance the professor in Arizona took his inspiration from Aldo Leopold, who often took students from his classrooms at the University of Wisconsin out on field trips, where he would then ask them to describe what they saw on the land. Why was there a clearing here? Why did a certain species of tree grow here and not another? Was the land healthy? Why or why not? Although Leopold was an advocate for beauty, as a scientist he wanted to know *why* it looked the way it did, what was happening, and how it was likely to change in the future. At the same time, art was necessary to interpret the facts that science provided. Of course, Leopold's own blend of literary and scientific skills is a powerful testament to this point.

Bill Zeedyk expressed a similar sentiment in his book *Let the Water Do the Work*, coauthored with Van Clothier. "Reading the landscape is both art and science," they wrote. "Science provides the facts; art interprets their meaning. A landscape artist sees form, pattern, texture, and tone, and then relates these to each other to form a harmonious composition. A scientist may view the same landscape and see not beauty, but function, and ask why." Why is the land formed as it is? Why is the texture coarse or smooth? Why is the tone light or dark? Why is this pattern here and that one there? Why does a particular kind of tree grow here, but not there? Why is the stream bank eroding? Is the composition harmonious or discordant? Why is the alluvium stratified as it is?

"The scientist is never satisfied solely with the beauty of the landscape for its own sake," they write. "She must question and probe until reaching an understanding of not only why the landscape looks as it does, but how it got that way and how it is likely to change in the future."

Beauty matters, but so does the question: what's going on here anyway?

As an example, let's take a walk together.

I'll start at the back fence of our property near Santa Fe, but before I take a step I'm going to ask myself a question (as you might): where do I live? I don't mean my city, county, or state, but rather the geography or landform of my home ground. I live on two acres of gently sloping cold desert dominated by juniper and piñon trees, bunchgrasses, and annual weeds. The elevation is 7,000 feet above sea level, and the annual precipitation over the past one hundred years has averaged 12 to 14 inches. The landforms around our house include small hills, arroyos, and tall mountains in the distance.

If I wanted to, I could do some research to answer a few questions before starting my walk: What is the geology of the land where I live? What are the actual soil types? What plants and animals might have existed here before humans began to make their mark? What were the historical uses of this land before it became a two-thousand-home subdivision spread out over 13,000 acres? How have precipitation patterns changed in recent years, and what effect has that had on the land and its life?

I know the soil here is easily eroded, and I've noticed that after a rain it *caps* in bare spots, meaning it forms a thin, hard crust. I know from experience that capping can be a problem if left undisturbed because crusts inhibit seeds from making it into the soil where they can germinate and grow. It also accelerates water runoff—and as I look around from our property line I can see a number of bare spots on the land.

Now we begin to walk—and as we do, let's pretend to be a raindrop for a moment. Everyone lives in a watershed—*everyone*—and thinking like a watershed is key to reading any landscape. All water that falls on the ground wants to go to the sea, thanks to gravity, so if you were a raindrop, how would you get there? In this case, our house sits near the tippy top of a small drainage, called a *greenbelt* here, which means it's easy to tell which direction the raindrop wants to flow: west. Water coming off our grassless backyard (thanks to our dogs and chickens) gathers together not far from the bottom of our property into a tiny stream that picks up other tiny streams as it goes. About a mile downstream it will merge with the Arroyo del Pueblo, coming in from the north, which merges eventually with the Galisteo River, to the south, and then on to a final merge with the mighty Rio Grande near Santo Domingo Pueblo, 30 miles from my house. Technically, we're part of the 460,000-acre Galisteo River watershed, but that's too much for this raindrop to comprehend, so we'll stick to our narrow greenbelt below my house.

The first thing I notice on our walk is that the footpath coming in from the right has captured most of the water flow in the greenbelt, causing it to become entrenched and as a consequence difficult to walk on. In many places, people have stepped off the footpath as they walk, creating a parallel track, which has begun to erode as well. On a steep section, I can see that water has spilled from the trail into the small channel, which is good, but I can also see grass *pedestaling* (grass plants that are confined to small pedestals of soil) and small rills in the soil on my right as I walk, which are signs of sheet erosion. I can see the sediment it creates deposited on the trail, where it is picked up by subsequent storms and carried downhill by rushing water, scouring the trail as it goes. I also know that storms in recent years have been much more intense than normal—and I can see the cumulative effects throughout the greenbelt.

This isn't an idle issue. There are no live streams in our subdivision, which means the 16 million gallons of water our community consumes each month on average during the summer must be pumped from underground aquifers, some shallow and some deep. The deep ones are comprised of fossil water, meaning it's been down there for a very long time, but the shallow aquifers are recharged by surface water, snow especially. The condition of the watershed directly influences the rates of recharge. If rain or snowmelt sinks into the soil, thanks to gentle gradients, decent vegetative cover, and uncapped soil, it boosts the recharge rate. If its shoots down eroded trails and washes off the land, however, the recharge rate drops significantly. In a high, cold desert that is enduring a low-grade but persistent drought, what's happening in our microwatersheds should be a concern to every resident.

My walk brings me to an intersection with another trail, where the greenbelt widens a bit. Near the junction are two tree stumps, both rough-cut by an axe, which suggests they were chopped down decades ago. It serves as a reminder that this area has been in use for a long time, first by prehistoric Native Americans, then by Spanish colonists, starting four centuries years ago, and now by the houses that line both sides of the greenbelt—including the dogs, cats, and people that live in them.

I keep walking. Eventually, I see a small headcut in the trail. Why is it there? Is it a matter of water velocity, a change in soil type, a small obstruction in the flow, or something else? I have my answer a short distance farther on when I see a much larger headcut—and up above it an electrical substation. This portion of the greenbelt is messed up. In fact, it looks like I'm standing on the edge of a shallow borrow pit, probably excavated to provide fill for the substation's base. Looking around, I see large chunks of cement and rebar among the weeds. Fortunately, the headcut doesn't show any sign of active erosion, so I keep going.

Shortly, I run smack into a wall of dirt—the elevated tracks of the Santa Fe Southern Railway. It's a spur that connects the main line of the historic Atchison, Topeka, and Santa Fe Railway at Lamy, 13 miles away, with downtown Santa Fe and is popular with tourists. When was it built, I wonder—1880? In any case, the elevated railway seriously disturbs the water flow here, channeling every drop into a 30-inch black pipe under the tracks. Looking closely, I see a line of debris on the railroad grade 2 feet above the pipe. The top of a recent flood! That means a pond was created in the greenbelt, which would explain the dense vegetation I see, including lots of purple aster, verbena, purslane, and kochia. There's even a stalk of corn! The pond was a convention of raindrops. I wonder what they discussed while waiting for their turn to shoot through the pipe?

I walk on.

I follow an old road that parallels the railroad track to the right, though it's essentially impassable to vehicles now due to the erosion. In fact, one section has become so deteriorated that someone official has installed a line of CAUTION tape on two wooden posts across it, possibly as a warning to the many bikers who zip along here. The tape makes it look like a crime scene, which in a sense it is. The old road rises gently to a divide between drainages and then drops quickly into Arroyo del Pueblo, the largest greenbelt in the subdivision—it's so large, in fact, that it merits a lovely wooden trestle for the train tracks. Peering at it, I see that a flood has cut 2 feet away from the bottom of the wash, exposing parts of the trestle's foundation, which can't be good. There's a large juniper tree jammed up against the structure as well. Hopefully the railroad people know all about it.

I turn right and head up the arroyo on a footpath. Shortly, I pass a ring of rocks on the ground that is filled with pottery, railroad spikes, and empty perfume bottles. It looks like a small, funky shrine, though to which deity isn't clear (an irreverent one, it would appear). As I keep going I begin to wonder if I'm walking on what used to be an old road. It's wide enough for a vehicle, but it's sunk 3 to 4 feet below the old floodplain, suggesting that if it had been a road at one point it caught the water flow and eroded downward. I know this is a quick way to alter the natural hydrology of arid environments, thanks to our easily erodible soils.

I keep going. The trail bobs and weaves around trees, rising out of the wash for a while, then slipping back down. Eventually the arroyo fans out, and I begin to see more traces of old roads. Fifty years ago, this country was an active cattle ranch, 13,000 acres strong, and if you look carefully you can see traces of old irrigation ditches on the land. Unfortunately, I can also see 2- and 3-foot headcuts in the old roads, each of which suddenly appeared three summers ago as the result of intense rainstorms. When I began walking in this arroyo ten years ago, none of the roads had a headcut. That tells a story of troubling changes in the region—but they were not the first sign of trouble.

As I walk, I also notice that long stretches of the arroyo's bottom rest well below the old floodplain, though this is not part of an old road, indicating that there have been serious erosion problems here in the past, likely the consequence of overgrazing by unmanaged livestock. It's another familiar story—too many cows, too little grass. There are three large earthen dams in the greenbelt—each constructed, I'm certain, in an attempt by the ranch owner to impound floodwater and slow the erosion. They probably doubled as stock tanks, but standing on them I definitely get the sense that humans were struggling to keep the land from unraveling here—and to some degree they succeeded. The arroyo's edges softened over the years and are now nicely vegetated, meaning they are no longer actively eroding. The dams didn't breach, which suggests the floods were kept under control, and the old roads were still in decent shape—at least until very recently.

As I near my exit point from the arroyo, I wonder what other questions I should be asking about the land as I walk. How about the plants that I see? What do the different species, age-class distribution, and vigor of the plants tell me? Does the area look like it's getting wetter or drier? Are there visible impacts caused by deer or other wildlife, besides the ubiquitous coyotes and rabbits? Has there been a fire recently? Any other natural disturbances? What other cultural impacts are there? What about the impacts of the modern roads? The mountain bike tracks I see? What about all the dog poop?!

I come to another headcut—a big one. What would we do if we wanted to repair it and restore the water and carbon cycles here? There are no wet meadows or riparian areas in this greenbelt that I know of (except behind the earth dams after a good rain), so carbon sequestration isn't really a possibility. The land could certainly grow more grass than it has, and thus store more carbon than it does currently, but that would be a tall order for a subdivision like ours. Most homeowners don't venture into the greenbelts, from my experience, much less try to "read" them from a land-health perspective. The headcuts don't threaten anyone's home (not yet), so there's no need to raise alarm bells. Still, reading the landscape on my walk tells me it could be in better shape—a lot better—if we wanted to make it so.

Do we?

A few years ago, the homeowners association hired a herd of goats, plus two handlers, to chow down on the weeds in the greenbelts, with great effect. The goats were popular, and for a while I felt optimistic about getting our greenbelts into better shape. Then came the chicken wars. Last year, an ugly row over backyard hens erupted in our community, dividing neighbors and effectively putting an end to the idea of using livestock to improve the land's health. It's a sign of both ignorance and illiteracy, I'm afraid. Not only do we not know how to read a landscape very well, but we've lost a sense for the positive role animals can play on the land. Mostly, we argue.

I don't want to think about that, however. It is a lovely day, full of clear, still air. As I step onto the road that takes me homeward, closing the circle of my walk, I think about wildflowers and possibilities, not arguments. There are a lot of metaphorical headcuts in the world today to go along with the literal ones, some of them huge. How will they ever get fixed? Fortunately, people like Craig Sponholtz and Bill Zeedyk give us hope, not to mention the rest of the folks I've met. An amazing, diverse, and effective toolbox has been pioneered and tested over the past thirty years, thanks to their hard work. We don't need to wait for something new under the sun to come along to restore the land, and ourselves, to health. It's here. We just need to learn how to recognize the signs, find a sweet spot, and get going.

The dirt road gives way to pavement and I turn left at an intersection. It's been a wonderful walk on a warm, late afternoon, and as I look up into the sky I see the clouds beginning to assemble themselves into another evening performance. The pinks and grays and oranges and maroons of our sunrises and sunsets remind me almost daily that the world is full of color, light, sound, touch, and other positive energy. It's an inspiring and hopeful time to be alive—if we choose to make it so. We can be rich. It also reminds me that we can't be spending all our time looking at our feet. We need to be looking up, at the clouds, at a world that is infinitely beautiful.

Seeing the Edible Forest for the Trees

ONE-TENTH OF AN ACRE IN HOLYOKE, MASSACHUSETTS

This is a story about two plant geeks, an urban sweet spot, and edible forests.

Before I start, however, I need to say a word about forests in general: I didn't look into them as I traveled around Carbon Country. Although forests play a huge role in the planet's carbon cycle and sequester a lot of atmospheric carbon dioxide as a result, I decided to skip green carbon. I did so for three reasons: First, forests have already received a great deal of attention from scientists, activists, regulators, authors, and journalists for their role in mitigating climate change. Soil carbon, in contrast, has received relatively little attention. Second, soils have a much greater carbon storage potential than forests (greater even than the atmosphere). In addition, soils can hold onto their carbon for longer periods of time—which brings me to the third reason.

Trees burn up.

During the summer of 2011, I watched a wildfire torch a forest from the back porch of my house in what became one of the largest conflagrations in New Mexico's history. Over 150,000 acres of woods went up in flames, cooking a large swath of soil as well. The fire released an unknown but probably impressive amount of carbon dioxide back into the air, transforming the forest from a carbon sink to a carbon source in a matter of hours. It wasn't an isolated occurrence. The general trend in the West is toward more frequent and larger fires, juiced by poor forest management in the past and hotter and drier conditions developing right now. As for the future, according to a forest expert I know, it isn't looking rosy. Many pine forests around the West are expected to convert to shrublands, thanks to repeated wildfires, which is rather depressing news. But that's the challenge with green carbon—trees die, get cut down, or burn up.

Edible forests, however, are a different story.

The two self-described plant geeks in this story are Eric Toensmeier and Jonathan Bates, and the edible forest garden they planted in 2004 resides on one-tenth of an acre behind a duplex home they bought in the rust belt city of Holyoke, Massachusetts. As a blurb in their book *Paradise Lot* puts it, their goal was to realize the urban permaculture garden of their dreams—and hopefully meet women to share it with them. However, the tiny property had a variety of serious problems: the backyard was lifeless; the soil full of brick and concrete bits; the narrow alleyways in deep shade; the steep, short front yard covered in asphalt; and the city's legal terrain hostile to composting, water harvesting, and livestock—chickens especially.[1]

It was perfect, in other words.

That's because Toensmeier and Bates wanted to see if they could bring a tiny spot of badly damaged land back to health by creating an *edible ecosystem* on it. That meant a forest garden, which is defined as an ecologically designed community of mutually beneficial perennial plants intended for human food production. Think fruits, nuts, berries, and certain veggies. Could they bring lifeless land back to life by gardening every square inch, they asked, creating a diverse and edible landscape? Would permaculture strategies developed in Australia work in the northeast United States? Could they grow banana plants in wintry western Massachusetts? If so, what else could they grow, and how could it serve as a role model for ecological restoration in cities using native perennial plants? Could their one-tenth-of-an-acre, in other words, yield big results?

The plant geeks decided to find out.

The route to Holyoke began in 1990 when Toensmeier, freshly graduated from high school, was bitten by the permaculture bug. He was intrigued by the basic permaculture equation: indigenous land management knowledge + ecological design + sustainable practices = landscapes that are more than the sum of their parts. He was particularly intrigued by its utility in designing food-producing ecosystems—but would it work in his native New England? No college at the time offered courses in this field of interest, so Toensmeier interned instead at various permaculture farms, eventually coming into contact with Dave Jacke, a forest garden expert. Together they organized an edible forest workshop in 1997, and Toensmeier never looked back.

Jacke is author of *Edible Forest Gardens*, a weighty two-volume how-to manual on edible ecosystems, written with Toensmeier. Jacke describes an edible forest garden as a perennial polyculture of multipurpose plants on a small plot of land that provide what he calls the seven Fs: food, fuel, fiber, fodder, fertilizer, "farmaceuticals," and fun. It's a forest, in other words, except it's a garden. He means it is gardening *like* a forest, not *in* a forest. Forest gardeners use the structure and function of a forest as a design strategy while adapting the design to meet human needs.

"We learn how forests work and then participate in the creation of an ecosystem in our backyards," Jacke wrote, "that can teach us things about ecology and ourselves while we eat our way through it."[2]

A forest garden is still a relatively novel idea, especially in America. The first temperate-climate home-scale forest garden was created by Robert Hart in 1981 on his tiny property in Shropshire, England. Hart's original goals, described in his pioneering 1991 book *Forest Gardening*, were self-sufficiency, home-grown nutrition, and minimal maintenance. During a visit in 1997, Jacke reported that the overstory in Hart's 3,800-square-foot garden included apple, plum, pear, elder, ash, and elm trees. The shrub layer included blackberry, gooseberry, raspberry, hazelnut, Japanese wineberry, Siberian pea, and white, black, and red currants. The understory of herbs included garlic mustard, mints, nettles, good King Henry, lemon balm, and sweet cicely. Although Hart didn't keep a record of yields, Jacke wrote, the garden clearly provided him with large quantities of diverse crops each year.

Hart also wanted to create a place that abounded in natural beauty. Although the garden didn't have the hallmarks of a typical

English garden, including bright colors, focal points, and a trimmed appearance, it had something else. "Its beauty derived from something deeper and more primal," Jacke wrote. "It shimmered with a special sort of energy. It felt safe, enclosing, and enfolding. . . . It felt like a forest, but it also felt like a garden."

What was missing from Hart's experiment, however, was a deliberate design based on ecological principles, carefully researched and arranged on the landscape to produce higher yields and greater diversity (and less leafy confusion). The ad hoc nature of Hart's gardening, including a lack of good paths, meant he had to work harder at harvesting and maintenance than he might have needed to otherwise, said Jacke. Still, Hart, who died in 2000, accomplished a great deal on his small plot, including providing the spark that ignited an edible ecosystem movement in temperate landscapes.

"Self-sufficiency, low maintenance, natural beauty, and decent-enough yields were successful products of this tiny piece of earth lightly cultivated by a humble, 'nonhorticultural' man," wrote Jacke. "Surely, this makes forest gardening a viable option within reach of millions."

A well-designed edible forest garden is a blend of art and science. To create one, a forest gardener needs to understand the perennial plant community, basic ecological processes, some organic chemistry, and some engineering—as well as employ a great deal of aesthetic intuition in order to make it all harmonize. Then there's the bigger picture. Forest gardens raise important questions, said Jacke. How do we live responsibly in the world? How do we grasp the interconnectedness of nature and our dependence on its webs of life in a culture and economy that value independence and reductionistic thinking? The answer, he said, is to look at the world through a different lens.

"The ultimate goal of forest gardening is not only the growing of crops," he wrote, "but also the cultivation and perfection of new ways of seeing, of thinking, and of acting in the world."

That's what Toensmeier and Bates were trying to accomplish behind their duplex.

The two friends met at a workshop, founded a seed company together, and before long hit on the idea of creating a food forest in the middle of a city. They knew from experience that the advantages of perennial plants, besides providing tasty food, included their ability to build soil, control erosion, improve rainfall capture, and sequester carbon. These could be very useful qualities in a blighted urban context, they thought. Also, there was another important advantage to perennials—minimal maintenance, which Toensmeier calls the "holy grail" of permaculture design.

"Having worked on annual vegetable operations and experienced the hard labor of planting and caring for annuals," he wrote in *Paradise Lot*, "I considered low-maintenance edible perennial vegetables an appealing alternative."

An affordable mortgage didn't hurt either.

The key to creating an edible ecosystem is a design that is as multifunctional as possible. To do this, Toensmeier and Bates spent an entire year observing and analyzing their one-tenth acre after they moved into the duplex in January 2004, contemplating their design. What part of the property received the most sunlight year-round (for the greenhouse)? What were the soils like, where was the best place for the pond, what guilds of plants would work best together in which part of the backyard?

Looking around the neighborhood for an ecological role model, they were delighted to discover a "feral landscape" behind a twenty-year-old Kmart shopping center. It was 10 acres of shrubs and wildflower meadows—perfect for their purposes. That's because nature was well on its way to healing the two-decades-old scar created by the development, and by studying the plants, they gained valuable clues to what nature likes to grow in a disturbed urban ecosystem.

"Most gardeners would not be excited about the species that were growing in the abandoned area behind the shopping center," Toensmeier wrote. "But to me, any plant community that can grow in such terrible conditions is a welcome one."

In 2005 they planted the garden. After sheet-mulching the bare ground behind the duplex (layers of straw, compost, organic fertilizers, and cardboard), they planted native persimmon, pawpaw, beach plum, clove currant, blueberries, juneberries, chinquapins (bush chestnuts), hog peanuts, grapes, pears, and the nonnative kiwifruit (but carrots and apples are likewise nonnative, Toensmeier noted).

In the front yard they planted banana trees.

By 2007 the garden was coming to life, a consequence of improving soils and the attractive habitat they had created for beneficial insects. The shrubs, perennials, and young trees were doing well, Toensmeier wrote, and the front yard already looked like a miniature tropical paradise. The banana trees, sheltered from westerly winds, collecting heat from the asphalt driveway, their roots protected from winter snows, became showstoppers in the area. Drivers stopped in the middle of the street to gawk. Puerto Rican neighbors asked permission to harvest leaves for tamales.

By 2009 their backyard ecosystem was showing "emergent properties," as they described it, meaning things were happening that were more than the sum of their parts. For example, they discovered a blue salamander under a 20-foot persimmon tree in the garden, which meant their edible ecosystem was attracting forest animals to patrol its understory—creatures that would never have survived in the yard in 2004.

Also emerging was love. By this time, both men had succeeded in their other principal goal: to find life partners. Children followed, as did chickens (once the local law was changed). Toensmeier wrote that cycling compost through the chickens made the soil so fertile that the top several inches of the garden were almost pure earthworm castings.

The forest garden, in other words, had taken on a life of its own.

In 2010 Bates kept a log of the amount and types of food coming into the kitchen from the garden. He estimated that, over a six-month period, they harvested 400 pounds of fruits and vegetables from this one-tenth acre, a total that was bound to rise in subsequent years as the edible ecosystem reached its full capacity. Best of all, the incredible yields were being produced with virtually no labor. It was a testament not only to the success of their design, but to the regenerative power of nature to produce life.

"The abundance in our garden comes to us in a self-renewing way," Bates wrote in *Paradise Lot*. "Our fruit trees are surrounded not by grass and asphalt, but by other useful and edible easy-to-care-for plants. After eight years, with very little care from us, all the plants are providing food, medicine, mulch, fodder, beauty, habitat, knowledge, seeds, and baby plants."

It was nothing short of amazing. "How is it that the abundance that I am now seeing in the garden," Jonathan exclaimed, "and in life, was hidden from me all this time?"

For Toensmeier, their little sweet spot demonstrated that cold-climate forest gardening *can* work. They created a multistoried forest garden in Massachusetts that can produce food from trees, shrubs, herbs, and fungi—even in the shade. They showed that ponds can grow food, asphalt can be a boon to tropical plants, and a good time can be had by all. There were challenges and setbacks (detailed in the book), of course, but after eight years they had accomplished everything on their original "to do" list, and more.

Toensmeier's next adventure, not coincidentally, is to write a book on carbon farming and tree crops. Not only do multifunctional perennial crops provide staple foods, they help build soil carbon, as he has written on his website.[3] They also offer ecological benefits like stabilizing slopes, rainwater harvesting, nitrogen fixation, and living fences.

Good stuff. What happened in Holyoke is another example of what's possible when we look at the land, and ourselves, in a regenerative light.

"While sustainability is focused on maintaining things as they are, regenerative land use actively improves and heals a site and its ecosystems," Toensmeier wrote in *Paradise Lot*. "Regenerative agriculture . . . achieves these goals while also meeting human needs. It's kind of an important topic for humanity this century."

6

EMERGENCE

I had never before climbed three flights of stairs to visit a farm.

That's what I did after leaving a subway station in Greenpoint, Brooklyn, and walking along Eagle Street to a warehouse owned by a television production company called Broadway Stages. In front of me were two gray metal doors set in a long redbrick wall. I stepped through one (feeling a bit like a secret agent) and climbed the stairs to the roof, where I saw hundreds of vegetables, green and growing, set in neat rows of dark, rich soil. Scattered about were the usual farm implements—hoes, hoses, rakes—and the occasional metal air vent. Walking to the edge of the building, I saw the serene East River below, and not far beyond it a sweeping view of midtown Manhattan and the Empire State Building.

Wow.

I had come to Eagle Street Farm to see the nation's first commercial rooftop farm in action and to learn about the role of rooftop farms in the rapidly expanding urban agriculture movement. I also wanted to meet Annie Novak, cofounder of Eagle Street Farm, and learn the story of how a Chicago girl who grew up reading *Vogue* magazine with dreams of "being fabulous" in New York City became a fabulous rooftop farmer. There was another reason to visit, and it was a big one: to try to understand *who is going to do the heavy lifting involved with the carbon work I had witnessed in my travels.* This work was just getting started, so I suspected it was going to get done by young agrarians like Dorn Cox, but I didn't know much more than that really. Who were these young people? What motivated them? What were their hopes and dreams?

I'll start with Annie's story.

It opens with visits she made as a little girl to an aunt on the Upper West Side. The charms of New York dazzled Annie, germinating a dream to live in the big city one day, a dream that blossomed into reality when she accepted an invitation to attend Sarah Lawrence College in nearby Bronxville to pursue a career in writing. Her mother was an artist and her father worked on the Chicago Board of Trade, dealing in corn and soybean futures. A life in agriculture wasn't part of Annie's plan.

This began to change in college when she became concerned about the negative effects of our industrial food system, especially the way it treated subsistence farmers in poor countries. This concern led Annie to focus on a subject dear to her heart: chocolate. Everyone loves chocolate, but practically no one knows how it is made or where the cocoa comes from, she realized. So she went to West Africa to find out. The experience opened her eyes not only to the living and working conditions of Ghanaian cocoa farmers, but also to the intense disconnect between food eaters and food producers. Our food system, she saw, had serious flaws.

However, it took a tragedy to propel Annie into a farming life. In 2005, shortly after graduation, her father died in an auto accident. Everyone in her family, all women, was faced with what she describes as a "moment of truth about our choices moving forward and how we were going to provide for ourselves."

Annie decided to become a farmer.

She started by learning how to grow vegetables. The week after graduating from college, she landed an internship that turned into a seasonal job at the New York Botanical Garden, teaching children how to grow food. In the years that followed, Annie balanced her city job with farming upstate, starting a nonprofit organization focused on young chefs, and dabbling in the restaurant business. During the winters, she traveled around the world to learn how to grow new crops and how to cook new foods, returning to New York City each spring to put her knowledge to work.

Through this career of patchwork jobs, Annie found her inspiration among the words of author and farmer Wendell Berry.

"I had been thinking of how difficult it is to grow in the city," she wrote in an article for the *Atlantic*, "of the lack of good soil and land access, of the smog and the trucks rumbling by on the Brooklyn-Queens Expressway. Sometimes it makes me happy, and other times monstrously depressed, to think of the stubbornness of nature below and above New York City: of weeds prying up through sidewalk cracks, or the red-tailed hawks circling their way back into our ecosystem by nesting on Fifth Avenue. How heartening, how necessary, that the passage I first opened to was this [by Berry]:

> In a country once forested, the young woodland remembers the old, a dreamer dreaming of an old holy book, an old set of instructions. And the soil under the grass is dreaming of a young forest. And under the pavement, the soil is dreaming of grass.

This proved to be the genesis of Eagle Street Rooftop Farm.

The project was the cooperative brainchild of Goode Green, a green roof company based in the city, and the owners of Broadway Stages, who supported the idea of growing green plants on top of their building and thus joining a global movement to rethink urban rooftops. By design, green roofs absorb stormwater runoff, reduce urban heat island effects, sequester carbon, improve air quality, increase biodiversity, and create a pleasant environment for people.

In 2009, workers with Goode Green hauled the necessary plastic sheeting, fiber filters, and 200,000 pounds of growing media to the roof of 44 Eagle Street and configured everything into a shallow field. Originally, the plan had been to create a nursery to grow sedum, a succulent plant popular with green roof owners, but Annie and cofounder Ben Flanner, who stepped forward as the first farmers, convinced the owners of the building to give veggies a chance instead. They added compost to the soil mix, planted crops they knew were tolerant to heat and water stress, organized a small cadre of volunteers, studied weather forecasts, and crossed their fingers.

It worked. Today, the farm grows a wide range of crops, specializing in heat-loving and dry-tolerant chiles. The farm also keeps bees, rabbits, and hens. It sells its produce on-site and to local restaurants.

It hasn't all been a bed of roses, however. As the first commercial rooftop venture in the city, the farm found securing the appropriate permits to be a major challenge, one that Annie says has been eased as urban agriculture gains traction in New York. Wind storms and unseasonable heat have bedeviled both the veggies and their handlers at times. Space is a limitation—she can't expand the farm even though she would very much like to. Fertilizer is another challenge, since it needs to be hauled up the stairs every time they need organic matter for the plants. This challenge has been partially solved by the introduction of rabbits and chickens to the farm, which Annie calls "my little poop machines." She would like to have a cow on the roof for the same reason, though that's unlikely.

Then there was the pollinator problem.

"Because there aren't a lot of green spaces in the city," she said, "we don't get a lot of pollinators, and that's a bigger issue than most people think." They resolved to start raising their own bees but quickly learned that Mayor Rudy Giuliani had made beekeeping illegal (putting bees in the same category as poisonous snakes). Eventually, new regulations were passed allowing apiaries, but by then a cold winter had killed off all of Eagle Street's bees. The farm responded by shifting to a cold-hardy Russian breed of bee.

The economics of rooftop farming are a challenge as well. The for-profit farm relies on value-added products like its hot sauce, called Awesome Sauce, to raise the $1.50 to $3 per square foot base value needed to farm unprocessed crops. At 6,000 square feet with no room to expand, farming at that scale makes just enough income to support a few part-timers, management included. But for impact far beyond its size, the economics are not as important as Eagle Street's educational purpose. And there Annie has found an eager audience.

"For folks who have daily nine-to-five jobs," she said, "it's nice to be able to come down on the weekends and get up to their elbows in dirt. One Sunday, all we did was carry up hundreds of garbage cans of soil to the roof. People were having a great time, spreading it like brownie mix. It's the hardest work you could think of, but people loved it."

Given the farm's small size, the most frequent question Annie gets is, "Can New York City feed itself?" Her response is unexpected: does New York want to? She thinks not. "The quality of our air and water is protected by upstate organic growers," she said. It's important to her that farmers, and the watershed in which they work, be supported by New York City residents. Not only are land and livelihoods protected this way, but an important urban-rural bridge is strengthened as well.

It became clear during my visit that, in addition to food and education, Eagle Street plays another huge role in the community: inspiration. Annie herself has become a role model for other young farmers. In addition to her food work, she has run in eight marathons, built and raced her own bicycle, made documentaries with the filmmakers of the Meerkat Media Collective, and acted in several films—and in her spare time she illustrates comics, makes stop-animation shorts using bits of felt fabric, and (of course) cooks for friends. She didn't drop her childhood goal of "being fabulous" either. In 2010, readers of the Huffington Post voted her the "Cutest Organic Farmer" in the country.

As for her own inspiration, after Wendell Berry she cites the intellectual challenges of farming. "In a world where we expect things instantly," Annie said, "agriculture is a lifetime of learning. There are very few other careers where you can continuously improve."

Then there's the inspiration of the city.

"At night, New York City is almost impossibly beautiful," she wrote in her *Atlantic* article. "Against the constellations of city lights and black sky . . . the small dark shadows of the chickens settling down to roost in their new coop were silhouetted by the brilliant white light of the Empire State Building. Finally, I saw what we were all looking forward to: the promise of new growth, the tenacity of plants, and the possibilities of city farming."[1]

Eagle Street has also inspired others to give rooftop farming a try.

In 2010, a group of young farmers formed a for-profit organization called the Brooklyn Grange and opened what has become the world's largest rooftop farm, located on two separate roofs in Brooklyn and Queens, totaling 2.5 acres (108,000 square feet). They grow more than 40,000 pounds of organic produce a year, with tomatoes being the biggest crop. Their goal is to create a fiscally sustainable model for urban agriculture while producing healthy food from what they call the "unused spaces of New York City."

"We believe that this city can be more sustainable," they wrote on their website, "that our air can be cooler and waterways can be cleaner. We believe that the 14% of our landfills comprised of food scraps should be converted into organic energy for our plants, and plants around the city via active compost programs."[2]

The work of Brooklyn Grange has quickly expanded to include egg-laying hens, a commercial apiary and bee-breeding program, and a farm training program for dozens of interns. The work also includes hosting thousands of New York City youths each season for tours and workshops, launching the New York City Honey Festival in 2011, and providing a unique setting for corporate retreats, dinner parties, and wedding ceremonies.

Like Annie, the folks at Brooklyn Grange believe that New York will always rely on rural farmers for the bulk of its food and that the relationship between urban and rural communities must be celebrated, but they also believe that having farms inside the city limits is an opportunity not to be missed. That's because urban agriculture can improve quality of life, create jobs, increase access to fresh foods, and provide educational opportunities to anyone who lives in and loves the city.

And it can tackle environmental challenges peculiar to metropolitan areas. With a grant from New York's Green Infrastructure Stormwater Management Initiative, Brooklyn Grange sited its second farm on the 65,000-square-foot roof of a building in the historic

Brooklyn Navy Yard, which allows it to manage over 1 million gallons of stormwater, reducing the amount that overflows into Brooklyn's open waterways.

There's another environmental benefit, though one less obvious to many: the 2.5 acres of soil under management of the Grange are soaking up atmospheric CO_2. It isn't much, of course, but it's a start—which raises a question: how many rooftop farms could New York City accommodate? No one knows the answer. It's an engineering question really: how many buildings can support hundreds of thousands of pounds of dirt on their roofs? Many, I bet.

In the meantime, rooftop farming continues to spread.

In 2013, it arrived in Boston with the launch of Higher Ground Farm, which occupies 40,000 square feet on top of the Boston Design Center, making it the world's second-largest rooftop operation. The brainchild of two young farmers, Courtney Hennessey and John Stoddard, the mission of Higher Ground is similar to what Annie Novak and the folks at Brooklyn Grange pioneered: make a dent in the urban heat island effect with a green roof; help with stormwater management; reduce carbon in the air; increase access to fresh, healthy food; create habitat for biodiversity; and provide educational opportunities, as well as many other cobenefits.

It's all part of an exciting movement to grow food within city limits. There are now gardens in city parks and schoolyards, farms in empty lots in Detroit, a vertical farm in Chicago, an aquaponics farm in Milwaukee, goats in Chicago, and on and on. After my visit to Eagle Street I tried to find out how many urban farms exist in the United States—without luck. I did learn, however, that there were twenty million "victory gardens" during World War II, providing over 40 percent of the nation's vegetables. I also learned that roughly 75 percent of the American population now lives in or near urban centers, which means the potential for urban farming is large, as is the potential for carbon sequestration in urban soils.

That's a lot of abundance waiting to happen.

"When I'm on a rooftop all I'm doing is listening to the sound inside a tiny seashell and trying to hear a larger ocean," Annie said. "If you live in a city, take advantage of it. Soak up the street smarts and the rush of city living that also embraces the outdoors and fresh tomatoes. You have to grow a small plot with a big picture in mind."

Spoken like a true agrarian.

At first blush, it might seem odd to call a fabulous rooftop farmer in Brooklyn who runs in marathons and acts in movies a young agrarian, but that's exactly what she is, and it is a sign of how young people are shaping the agrarian ideal to fit a changing world. Think of Annie as you read this quote from Wendell Berry:

> The agrarian population among us is growing, and by no means is it made up merely of some farmers and some country people. It includes urban gardeners, urban consumers who are buying food from local farmers, organizers of local food economies, consumers who have grown doubtful of the healthfulness, the trustworthiness, and the dependability of the corporate food system—people, in other words, who understand what it means to be landless.[3]

What does *agrarian* mean, exactly? In Latin, it means "pertaining to land." My dictionary defines it as relating to fields and their tenure or to farmers and their way of life. Wendell broadens this definition, calling it a way of *thought* based on land—a set of practices and attitudes, a loyalty and a passion. It is simultaneously a culture and an economy, he says, both of which are inescapably *local*—local nature and local people combined into "a practical and enduring harmony." The antithesis of agrarianism is industrialism, which Wendell says is a way of thought based on capital and technology, not nature. Industrialism is an economy first and foremost, and if it has any culture it is "an accidental by-product of the ubiquitous effort to sell unnecessary products for more than they are worth."

An agrarian economy, in contrast, rises up from the soils, fields, woods, streams, rangelands, hills, mountains, backyards, and rooftops. It embraces the coexistences and interrelationships that form the heart of resilient local communities and local watersheds.

It fits the farming to the farm and the forestry to the forest, to which I would include the meander to the stream and the carbon to the carbon cycle. For Wendell, the agrarian mind is not regional, national, or global but local. It must know intimately the local plants and animals and local soils; it must know local possibilities and impossibilities. It insists that we should not begin work until we have looked and seen where we are; it knows that nature is the "pattern-maker for the human use of the earth," as he describes it, and that we should honor nature not only as our mother, but as our teacher and judge.

Even at the top of three flights of stairs.

And under the pavement, the soil is dreaming of grass.

A day later, I traveled up New York's Hudson Valley to visit a young leader of the emerging agrarian movement. She has the aristocratic-sounding name of Severine von Tscharner Fleming, and when I asked for directions to the farm she replied, "Third mansion on the left." I thought she was joking. Nope. I passed FDR's home first, then the Vanderbilt estate, followed by the Mills mansion, where I turned off from the main road. Wow. Partly a state park, the Mills lawns were spacious, green, and very tidy. I spied the stately mansion through the trees. I knew the Hudson River was just ahead. Maybe Severine *was* joking. A farm, here? I wasn't sure until I pulled up to a weather-beaten farmhouse near the river and saw a funky bicycle leaning casually against a wall. It had to be Severine's. A few moments later she burst outside and waved a greeting. A self-described "punk farmer," Severine has hair as unruly as her politics—and yet here she was, planning an organic farm on groomed grounds once owned by a former US Secretary of the Treasury.

Somehow, it all seemed appropriate.

We went for a walk to see her latest project. I had met Severine a few times before, and I knew her to be an astonishingly energetic and successful advocate for young farmers like herself. For starters, in 2007 she founded the Greenhorns, a nonprofit that has become an influential grassroots network dedicated to recruiting and supporting young farmers and ranchers. Like its director, it is not a shy organization. "America needs more young farmers and more young farmers want a piece of America," is how the Greenhorns website describes its mission.[4] This didn't simply mean access to a piece of land either, though that's a huge issue for young farmers today.

We walked into a dilapidated greenhouse on the edge of an old farm field. As is typically the case, advocacy work doesn't pay many bills, so Severine signed a contract with the owner of this portion of the old Mills estate, a wealthy hotelier, to design and implement an organic farm on the property, which will include veggies, herbs, flowers, Muscovy ducks, chickens, rabbits, and possibly pigs. I was impressed—it looked like a daunting job! The grounds had been farmed in the past, but long ago. It was a mess now. A large portion of the greenhouse had collapsed and the farm's infrastructure was in need of serious upgrades. Severine was up for the challenge, of course.

It isn't just her goals that are impressive, but also the variety of means by which she accomplishes them. She calls it "avant-garde programming," and it includes videos, podcasts, e-books, and Web content, naturally enough, but it also includes workshops, social mixers, barn dances, art projects, and a full-length documentary, all done in a bouncy style that can only be described as "farm-hipster."

If Greenhorns wasn't enough, Severine also cofounded the National Young Farmers Coalition, manages a weekly radio show on Heritage Radio Network, writes a popular blog, speaks at countless conferences, organizes endlessly via the Web, and helps with something called the Seed Circus, which puts on educational events for young farmers all around the country. *And* she's assisting the Schumacher Center for a New Economics, located just over the Massachusetts state line, with an initiative called the Agrarian Trust, which aims to help young farmers gain access to land. *And* she was editor-in-chief of the 2013 *New Farmer's Almanac, and* she has built an eight-thousand-volume agricultural library.

By the way, she's a farmer too. At her previous gig, on Smithereen Farm in Essex, New York, she helped produce organic pork, rabbit, goose, duck, culinary herbs, teas, and something called "wildcrafted" seaweed. She's also been active in local farmer's markets, selling produce. It all sounds exhausting and exhilarating.

"The young farmer movement looks and sounds romantic, and it is," Severine said. "It also is ridiculously difficult to break into farming these days. And it is critical that we do so. People who take on this challenge are highly tenacious, ambitious, inventive, and also either stubborn or a little nuts."

Words that could easily describe Severine herself.

The underlying theory of her advocacy work is that many more people would choose to farm if they knew how to get started. She also wants young farmers to understand that it's possible to have a social life *and* a viable business at the same time and, once they've begun farming, to make sure there's a network of support to help them get access to the resources and information they need to stay in business.

"Basically, if the young farmer makes it beyond the third year and still loves it," she wrote in an article for Quivira, "they'll likely stay a farmer for life. Which is of course what our country desperately needs. USDA says it wants to bring on 100,000 new farmers in the next five years. It's a big project, and yes I think art, culture, free beer, delicious food, hot sexy farmer men, and sweaty dancing are appropriate recruitment tools, far more effective, in the long run, than government-issued propaganda."[5]

Possibly describing her own proclivities, Severine says young people are inspired to get into farming for both political and environmental reasons. It starts typically with a journey through apprenticeships and internships as each young farmer discovers which aspects of the farming life he or she wishes to pursue, followed by a bunch of hard work to gain proficiency in, say, carpentry, horse wrangling, or irrigation system maintenance—without going into debt, and usually before starting a family.

Who are these young farmers? According to Severine, most are from cities and suburbs—thus the "greenhorn" moniker—and many come from the social justice or food poverty movements. Another portal is the FoodCorps, which is a project of AmeriCorps and places young people in food-oriented jobs, often building school gardens. Many young farmers attended farm camps when they were kids or went on field trips to local farms through their elementary schools. A few participated in 4-H, though not as many as you might think. The educational backgrounds of young farmers today vary widely, including engineering, public health, computer science, literature, anthropology, and earth science, but the decision to go into farming after examining all the options is based on the same principle: to live a life with dignity and purpose and have a positive impact on the community.

Sounds like agrarianism to me. As do the challenges.

"We'll seize opportunities to buy inexpensive battered pastures and compacted soils," she wrote, "and then heal those lands using good land stewardship techniques. We'll reclaim territory from commodity crops, and try our best not to churn or ruin our own soils while we build up enough capital to stop roto-tilling. We'll process our own darn chickens and build our own darn websites. We are just as stubborn and innovative as farmers have always been."

Severine's own journey into an agrarian life began with childhood visits to her mother's farm, which had been in the family for six generations before being lost. Severine spent summers hanging out with cows and kittens, as she described it, and playing in the hayloft and investigating bugs and eating cherries and getting stung by bees. It was an experience that stayed with her all the way to Pomona College. After a transfer to the University of California–Berkeley, where she graduated in 2008 with a B.S. in conservation and agroecology, Severine founded the Greenhorns. Its genesis came about unexpectedly. Helping to organize a film festival on food, Severine went looking for films on sustainable agriculture to inspire her fellow students—and found none.

"When I went to the basement of UC Berkeley Library to find films for the film festival, what I found was a doom-dominated media landscape," she said in an interview. "Horror movies about soil erosion, slave labor in sugarcane, soybean plantations displacing rain forest, persistent hunger, the dust bowl migration, et cetera It was depressing."

The experience inspired her not only to start Greenhorns, but to begin filming young farmers and their positive stories. Encouraging young people to go into farming, she knew, required an upbeat message. It also required a twenty-first-century toolkit: new media, new formats, and radical ways of sharing information and resources, including crowd sourcing, community investment funding opportunities, and open-source sites like Farm Hack. At the same time, tried-and-true strategies are necessary as well, she said, such as adaptive management and holistic decision-making models. She also realized that collaboration is often a "winning business tactic." While computers are important, most young farmers recognize that for a business rooted in place, chatting about the weather and passing the time of day with neighbors are still some of the most effective tools they have.

Beyond the practical and the social, there's a third arena that young farmers must occupy, Severine insisted. If they are going to be successful in the long run, politics and policies have to be addressed—confronted, actually. To this end, she cofounded the National Young Farmers Coalition, a network of young people and like-minded organizations fighting for needed policy changes, including shifts in federal programs via the Farm Bill. Changes proposed by the coalition include government investment in local food system infrastructure, low-interest loans for beginning farmers, training programs, funding for greenhouses, and incentives for landowners to lease their land to beginning farmers.

It's all laid out in legislation called the Beginning Farmer and Rancher Opportunity Act, which was introduced in Congress in 2013 as House Bill 1727 and Senate Bill 837. Alas, it hasn't made much progress, and Severine is quick to admit that changing farm policy in the current partisan political climate in Washington is a tall order.

She's equally sober about the other challenges confronting young agrarians.

"As the saying goes, the closer you get to the ground, the denser the weeds," she wrote. "Listen to any discussion between farmers, old or young, and you'll hear about the heartbreak and the challenges to stay on or obtain land in a highly distorted agricultural economy. Farming involves never-ending facilities upkeep, liability insurance, parking tickets at the farmer's market, food-safety regulations, health insurance costs, and the dysfunctional remnants of a threadbare farm service sector."

But there is good news too, she pointed out. According to the USDA Census of Agriculture, the number of young people farming in the United States is on the rise. Though it is still a tiny minority of the tiny minority of Americans who are farmers, it reinforces the argument that a movement is growing.

"Big things start small and those of us in this new farmers' movement are still running small or medium-sized operations, gaining experience and knowledge and aching to scale up," she wrote at the end of her article. "We could do more with just a little help, a special chance on a piece of land, a great deal on equipment, babysitting, help with accounting, a graphic design tip, or low cost advice from an attorney. We will continue to need mentorship and guidance, and the occasional kick in the pants. It will be hard, but it will not be boring. Don't forget that we may need a pep talk every now and again."

I bet.

I've wondered at times how young people keep their spirits up. When I was a twentysomething, the world just wasn't as complicated and daunting as it is today. That's a truism, of course, but it's also true. I can't imagine, for example, what it would be like to come of age under the specter of climate change or knowing there will be nine billion people to feed in 2050. Our political system wasn't broken either, not like it is now, and the long arm of corporate capitalism hadn't yet insinuated itself into all aspects of our culture and economy. True, young people today have envious amounts of marvelous technology at their disposal (my college graduating class was the last one to write its senior theses on a typewriter!), but this development has also added to the frenetic pace and complexity of modern life, not lessened them, as promised by the early advocates of personal computers. Keeping spirits up can be difficult in the face of all this. Fortunately, farmers like Severine and Annie are part of an emerging cohort of like-minded young people determined to make a difference in the world—a cohort that is part of an emerging movement that I think is very hopeful.

And the name of this movement is new agrarianism.

I first ran across the term in 2003 in a book of essays on the topic collected and edited by Eric Freyfogle, a law professor at the University of Illinois. The book's title, *New Agrarianism*, resonated with me because it described exactly what I was seeing on the land as I traveled around for Quivira. In fact, I could have used Freyfogle's own words from his essay, "A Durable Scale," to describe my experience. "Within the conservation movement," he wrote, "the New Agrarianism offers useful guiding images of humans living and working on land in ways that can last. In related reform movements, it can supply ideas to help rebuild communities and foster greater virtue. In all settings, agrarian practices can stimulate hope for more joyful living, healthier families, and more contented, centered lives."[6]

His words, I'm afraid, didn't impress my colleagues in the conservation movement at the time. Nevertheless, in his essay Freyfogle produced a list of new agrarians that was spot on:

- the community-supported agriculture group that links local food buyers and food growers into a partnership, one that sustains farmers economically, promotes ecologically sound farm practices, and gives city dwellers a known source of wholesome food
- the woodlot owner who develops a sustainable harvesting plan for his timber, aiding the local economy while maintaining a biologically diverse forest
- the citizen-led, locally based watershed restoration effort that promotes land uses consistent with a river's overall health and beauty
- the individual family, rural or suburban, that meets its food needs largely through gardens and orchards, on its own land or on shared neighborhood plots, attempting always to aid wildlife and enhance the soil
- the farmer who radically reduces a farm's chemical use, cuts back subsurface drainage, diversifies crops and rotations, and carefully tailors farm practices to suit the land
- the family—urban, suburban, or rural—that embraces new modes of living to reduce its overall consumption, to integrate its work and leisure in harmonious ways, and to add substance to its ties with neighbors
- the artist who helps residents connect aesthetically to surrounding lands

- the faith-driven religious group that takes seriously, in practical ways, its duty to nourish and care for its natural inheritance
- the motivated citizens everywhere who, alone and in concert, work to build stable, sustainable urban neighborhoods, to repair blighted ditches, to stimulate government practices that conserve lands and enhance lives, and in dozens of other ways to translate agrarian values into daily life

To this list I could add only the following:

- the carbon farmer or rancher who explores and shares strategies that sequester CO_2 in soils and plants, reduce greenhouse gas emissions, and produce cobenefits that build ecological and economic resilience in local landscapes

Freyfogle shares Wendell Berry's belief that agrarianism is the proper countervailing force to industrialism and its surfeit of sins, including water pollution, soil loss, resource consumption, and the radical disruption of plant and wildlife populations. Freyfogle goes on to add broader anxieties: the declining sense of community; the separation of work and leisure; the shoddiness of mass-produced goods; the decline of the household economy; the alienation of children from the natural world; the fragmentation of neighborhoods and communities; and a gnawing dissatisfaction with core aspects of our modern culture, particularly the hedonistic, self-centered values and perspectives that control so much of our lives now.

In contrast to these negative attributes of modern life, the new agrarianism is first and foremost about living a life of positive energy and joy, says Freyfogle. Nature is the foundation of this joy, but so are the skills necessary to live a life. At its best the agrarian life is an integrated whole, with work and leisure mixed together, undertaken under healthful conditions and surrounded by family.

"When all the pieces of the agrarian life come together," Freyfogle wrote, "nutrition and health, beauty, leisure, manners and morals, satisfying labor, economic security, family and neighbors, and a spiritual peacefulness—we have what agrarians define as the good life."

It is to this definition of the good life to which Annie Novak and Severine and the other new agrarians strive.

At the same time, it's very important that the new agrarianism not forget the positive qualities of the *old* agrarianism.

This is one of the lessons I take from the work of Miguel Santistevan, a youthful and dynamic teacher, farmer, activist, and lecturer based in Taos, New Mexico. Miguel is dedicated to mentoring young people in sustainable (i.e., traditional) agriculture, reminding them through hands-on experience that what's new is actually quite old.

Miguel, born and raised in northern New Mexico, has an impressive list of credentials: a master's degree in ecology from the University of California–Davis; a Ph.D. candidate in biology at the University of New Mexico; certified in permaculture; a former high school science teacher; director of youth-in-agriculture programs; public radio show producer; *mayordomo* for a local *acequia* (irrigation ditch) system; heirloom seed saver; farmer on his family's land; nonprofit founder; and father and husband to boot! The tie that binds all of these activities together is Miguel's deep attachment to the centuries-old acequia-irrigated and native dryland agricultural systems of the Upper Rio Grande region.

The old agrarianism—facing modern challenges.

"Our youth are our seeds," Miguel said at a conference on young agrarians that I helped organize a few years ago. "But what kind of conditions are we creating for them to germinate into the future?"[7]

To answer his question, Miguel works principally on two fronts. First, every year he teaches groups of high school–age youths how to grow food according to New Mexican traditions, often on as many as four farms simultaneously. He mentors them through the entire process of picking seeds, planting them, nurturing the crop, and harvesting it, emphasizing age-old staples such as corn, beans, squash, chiles, and melons. He also educates them in the four-hundred-year-old culture and history of family-scale agricultural production in the region, including the vital role of acequias—an institution with North African roots that is as much about governance and communal sharing of a precious resource as it is about the mechanics of getting water from a river to the farm field.

Second, his research focuses on locally adapted heirloom varieties of crops, called *landraces* by botanists, that have been maintained in particular communities or regions for generations. Specifically, he's interested in plants that have proven to be resilient to adverse climatic conditions—landraces that have been associated with their local soils and waters (and farmers) for so long that they have practically seen it all: late frosts, hailstorms, drought, floods, pest attack, competition with weeds, and early frosts. Over the years, these events have "thinned the herd" of the genetic base of the crops, as he put it, leaving only the most resilient members to propagate future generations. It's an ancient practice: by saving the seeds of resilient parent plants, traditional farmers around the world can cultivate adaptable strains of crops that grow well under a wide range of difficult climatic conditions.

This isn't merely an academic interest, as you can guess. Miguel believes these landraces are key to agriculture's future in the era of climate change. That's not speculation either—he and his young farmers know so from their own hard experience.

In 2011, the snowpack in the mountains above Taos was very light, causing the acequias that fed the farm fields to go dry by the third week in June. Since Miguel and his students don't use water from underground wells, they had to rely on old-fashioned technology to keep their crops alive: watching the clouds (and paying regular visits to the Weather Channel). No substantial rains arrived, however. Instead, it sprinkled rain now and then, and a few drops made their way down the plant stems to the roots, but that was all.

It was enough, as it turned out.

"Our *alberjon* [peas] were flowering and were able to set seed in the next several weeks without substantial water," Miguel said in a return talk at Quivira's annual conference, "The *maíz blanco* [white corn] looked shorter than usual with water-stress-afflicted leaves, but when it was all harvested, we had several ears of corn that were obviously unaffected by water shortage, confirming all my beliefs in the potential resilience of this ancestral staple. Some other crops, such as lentils and fava beans, shriveled up in the heat as if they were burned under a magnifying glass. Surprisingly, we were still able to bring in a few dozen seeds of each."[8]

The real challenge in weathering the drought, as it turned out, was not the persistence of the crops but rather the persistence of herbivores. As the dry weather took a toll on the crops, it also took a toll on nearby wildlife—who began to eye the crops hungrily. The lack of a constant human presence in the fields, due to logistics and obligations, gave the critters the break they needed. Prairie dogs obliterated one farm field completely and magpies decimated the corn in another field. A friend of Miguel's reported that elk had wiped out her entire farm. The experience forced Miguel and his student farmers to look hard at the challenges of reinvigorating the local food system under the stress of uncertainty caused by climate change.

It also caused Miguel to fret. Had this experience diminished the enthusiasm of the young farmers, who had literally watched the fruits of their labor disappear week by week? He understood the inherent gamble in agriculture, especially in the context of climate change, but how would the students handle the disappointment? Fortunately, they reassured him they were looking forward to the next round of planting activities. And they expressed their concerns and hopes in the short videos and podcasts that Miguel encourages each student to make as part of their learning experience. In the end, Miguel realized that the year's *real* harvest was coming to grips with the extent of the challenge that lies ahead.

"I am reminded of a *dicho* [saying] that is common to our region," he said, "that we plant three seeds in each planting place *para mi, para vos, y para los animalitos de Dios*—for me, for us, and for all of the animals of God. This saying reflects much wisdom, but was likely developed at a time when the effects of suffering wildlife could be spread out over a larger area during years of drought. We will have to be innovative about how to secure a harvest for us human beings while also providing for *los animalitos de Dios* in a way that brings about more balance to our ecosystems and for all the organisms therein."

For all the challenges facing the Southwest and its food producers, including predictions of hotter and drier conditions, Miguel remains optimistic. That's because the region is home to cultures and crops that are well adapted to extreme and uncertain climate patterns. Miguel's goal is to develop agricultural practices that honor the experience and capability of the crops and cultures inherited from past generations.

"We never use a prediction of drought to deter our plans of planting," he said, "rather, we embrace the difficulties as an opportunity to discover the 'champions' in our crop populations while we hone our techniques to learn how to meet the challenges of drought and climate change."

One example of a traditional sweet spot in the region that has stood the test of time is a *milpa*: an unplowed plot of farmland planted with corn, beans, and squash—often referred to as the Three Sisters. As native tribes have known for centuries, these three crops are complementary, both ecologically and nutritionally. Corn (maize) requires high levels of nitrogen in the soil to grow properly. Bean plants fix nitrogen in the soil, and the maize repays the debt to the beans by providing stalks for the bean plants to climb. Squash grows between the corn rows, providing mulch and helping to keep the weeds down. The edges of many milpas are usually planted with chiles, as a pest control, interspersed with melons and watermelons. In this way, space is saved, water conserved, and labor reduced. Milpas can be fertilized by compost or other organic material, such as ashes from kitchen fires and manure, enriching the soil. On tiny amounts of land, the bounty can be plentiful.

It also sounds like a good way to store a bunch of carbon.

In tropical Mesoamerica, by the way, milpas are an important part of subsistence (sustainable) agriculture, and have been since the days of the Maya and Aztecs. There are four stages to a typical milpa, spread out over twenty to thirty years. I'll cover it briefly here as a demonstration of another type of sweet spot:

1. The tropical forest is cleared, burned, and planted. For a few years, the Three Sisters grow in full sun. Below them are herbs, tubers, and other plants cultivated by the farmer to reduce pests and enhance the soil.

2. The milpa evolves into a forest garden. Quick-yielding fruit trees, such as plantain, banana, and papaya, are planted among the Three Sisters, followed by avocado, mango, citrus, and guava trees.
3. The fruit trees mature, producing fruit and creating a new canopy, blocking the sun and inhibiting undergrowth. The Three Sisters go out of business. Hardwood trees, such as cedar and mahogany, are planted.
4. The forest garden becomes a hardwood forest. The farmer lets the hardwoods grow tall and create a high canopy. Eventually the farmer will harvest them, then clear and burn the forest and start a new milpa all over again.

There's a lot more to a milpa than this, of course, including complex social and cultural relationships, but basically it's a sustainable way to maintain wildlife habitat while producing plants for food, spice, shelter, medicine, ornaments, and profit. One professor called it "one of the most successful human inventions ever created."

Miguel is quick to give credit to indigenous peoples for innovating these types of practices. He understands—as we all should—that we're standing on the shoulders of those who went before us. In fact, there's a phrase used by Native Americans for the type of farming that Miguel advocates: "organic and sustainable by tradition." Here's a list that I found on a First Nations website—see if you can tell which are part of the old agrarianism and which are part of the new:

- food is produced in concert with nature
- appropriate technology and scale
- complemented by hunting and gathering wild food
- concern for biodiversity
- family and community activity
- agriculture is part of spiritual and cultural values
- age-old practices and experience, especially with drought
- sufficiency and cultural survival are goals, as well as profitability

- inclined toward limits to growth
- not a traditional dependence on nonrenewable energy sources
- reciprocity is important
- locally adapted seeds and seed-saving are traditions
- regenerative economy is a goal
- food security: a community is described as "food secure" if all its members have access to nutritionally good, safe, culturally acceptable food at all times

Interwoven with the traditional practices that Miguel studies and employs is an important cultural idea called *querencia,* a term that has linguistic roots that stretch back at least to 1611. According to historian Juan Estevan Arellano, a milpa farmer himself and an important mentor to Miguel, *querencia* means "anchored to a certain place" and "the inclination or tendency of man and certain animals to return to the site where they were raised." It's a place of rest and safety, a place *conoce como sus manos*—you know like your hands. A place that shelters the heart of agrarianism.

"*Querencia* is the ethic behind how we look at the land and water," Arellano said at a Quivira conference, "and when you love something with so much heart, *con tanto corazón,* then you are going to take care of it."

Querencia involves affection, longing, and a sense of responsibility. It can refer to a memorable time of day, a kind of weather, a favorite music, food, taste, or smell. Think of red chile sauce or green chile stew—and you think of New Mexico. It also refers to who you are and what you do best. For Miguel, *querencia* is both the land around his home in Taos and his desire to teach, mentor, and inspire.

"Ask yourself," said Arellano, "where do I come from? Where do I feel most at home? As a writer, my desire to write is my *querencia*."[9]

What's your *querencia*?

Bryce Andrews finds his *querencia* on the ranches of Montana, and in many ways his story sums up the challenges and the hopefulness of the new agrarianism.

Bryce was born and raised in Seattle, the son of a professional photographer and the director of the University of Washington's art museum. Normally, this would suggest a *querencia* of gray skies and rainy streets, but Bryce's life took a fateful turn at age of seven when he and his parents drove to Billings, Montana, for a vacation stay on the ranch of sculptor Pat Zentz and his family. Bryce spent most of his time pulling weeds, as he recalled, while his mother photographed every skeletal cottonwood and deteriorating outbuilding she could find on the ranch. He also recalled the impression the vastness of the sky and the stars at night made upon him.

He cried when they headed home.

Coincidentally, his father had opened a new exhibit at the museum titled *The Myth of the West*, which fascinated the impressionable youth. Bitten, Bryce returned to the Zentz ranch the following summer—and every summer after that until he turned eighteen. They put him to work fixing fences, moving cattle, and a hundred other ranch chores, all for essentially no money. He loved every moment and took his pay in the pleasure of being in Montana's wide-open spaces. "When the work was done I lay faceup on the truck's roof looking into the deep blue bowl of the sky," he told the audience at the same conference that Miguel spoke at. "Whenever I went home to the damp claustrophobia of Seattle, I would dream about big, dry, lonely country."

His *querencia*.

In 2006, Bryce returned to Montana for a seasonal job on the 25,000-acre Sun Ranch, which is located in the upper end of the beautiful and biologically rich Madison Valley, northwest of Yellowstone National Park. On one side of the Sun was the Lee Metcalf Wilderness, located on US Forest Service land and primo habitat for grizzly bears, black bears, elk, wolves, lynx, and wolverines. On the other side was the fly-fishing mecca of the Madison River, primo habitat for tourists, sportsmen, and wealthy second-home owners. As for the job itself, the position title was "Assistant Grazing Coordinator," but what caught Bryce's attention were three of the required skills listed in the job description: common sense, adaptability, and gumption. He knew the Sun was in the vanguard of a movement to rethink agriculture in the West, so he jumped at the opportunity.

The Sun *was* an unusual ranch, as I can attest. I visited it twice (both prior to Bryce's arrival) and wrote a profile of its innovative manager, Todd Graham, and the conservation ranching practices he directed on the property. Its previous owner was a Hollywood movie star, and when the ranch changed hands in 1998, the new owner had a different mission in mind: to find a way for wildlife—elk in particular—to coexist with livestock. Todd began by "freshening up" the old, gray grass on the property for the elk using the tools of cattle grazing and electric fencing on portable posts. He created a sequence of five 100-acre pastures on the ranch and then turned out a portion of the ranch's 1,300-head cattle herd for six to eight days in each pasture early in the summer. The cattle ate the old grass along with the new, reinvigorating both, and when the grazing rotation was done the fence was rolled up and removed—leaving not a trace.

It worked well, ecologically and economically, as I saw.

Coexistence between cattle and wolves, however, proved far more elusive, as Bryce quickly found out. When a pack of wolves denned on the ranch during that first summer and began to kill heifers, despite strenuous efforts by Todd and the staff to avoid a conflict, the order came down to shoot two of the wild beasts. For Bryce, simultaneously a city-raised conservationist and a livestock-raising ranch hand, it was a difficult call.

What happened next, however, was shattering: he shot and killed a wolf himself.

"I could not stop seeing the kill," he wrote in a memoir called *Badluck Way*. "The wolf emerged again and again from the trees. Each time, I shouldered the rifle and squeezed the trigger, living again the explosion, the impact, and the ringing silence that followed."

His experience continued to haunt him:

> How do you get over something like a wolf? You don't, really. Working like a madman helps. I herded and settled cattle that the wolves had scattered. I watched diligently for signs of infection and disease. I slept out nights and listened for howling that never came. Having exhausted the ranch's supply

> of broken fences, I spent large chunks of time in the saddle and worked on my horsemanship and herding skills in earnest.
>
> I thought often of the wolves and could not blame them for the wrecks of summer. We had brought the cattle to them, after all. We bred animals for meat and docility and then dropped them on the doorstop of the howling wild. As we did it, we talked about shoehorning livestock into ecological niches of wild grazers, the cattle functioning as part of the ecosystem. I decided that we had been a bit too successful. We had moved our stock across the land like so many buffalo or elk, and the wolves had taken notice.
>
> I lost a lot of sleep over the wolf I killed, worrying endlessly over whether my actions were right, just, or ethical. I never came to a lasting conclusion. It seemed more germane and natural to say that the killing was necessary, unavoidable, and unfortunate, and then move on.[10]

Bryce decided to move on. In 2007, the owner of the Sun Ranch, a wealthy Silicon Valley entrepreneur, decided to subdivide a portion of the ranch and sell the housing lots to other millionaires to pay mounting bills, a decision that changed the conservation purpose of the property and broke a promise he had made to his employees. In short order, Todd, Bryce, and others left, never to return. In 2010, the Sun itself was sold to the CEO of a mining corporation for many millions of dollars, putting an end to an important role model of conservation ranching in the West, breaking a lot of other hearts as well, mine included.

We push on, however.

After a stint at Montana State, where he earned a graduate degree in environmental studies, Bryce's career took another new agrarian turn when he was hired by the Clark Fork Coalition, a decades-old environmental advocacy organization based in Missoula, to manage its newly acquired Dry Cottonwood Creek Ranch, located on the Clark Fork River near Deer Lodge. It was an important sign of the times: an environmental group that had been dedicated to fighting the copper industry and forcing it to clean up the tremendous damage it had done to the river was now in the ranching business. The coalition's motivation was straightforward: to protect the open space along the river from subdivisions. That meant supporting ranchers—and getting into the beef business as well. It was all part of the burgeoning collaborative conservation movement at the time, and Bryce found himself right in the middle of it.

It began with a visit by his class on environmental restoration to the toxic disaster zone that was the upper Clark Fork River, site of the largest Superfund cleanup project in the nation at the time. After visiting the yawning maw of the Berkeley Pit in Butte, and its purple lake of acid, they drove to the defunct mining mill complex at Anaconda, ground zero for a century's worth of copper mining, smelting, and dumping. The site bled toxic waste like an open sore. It was shock, Bryce said, but not quite like the shock they felt further downstream, when they saw their first *slickens*.

"For the uninitiated," Bryce recounted in an essay for the Quivira Coalition, "there is nothing quite like the first encounter with these bizarre dead zones. Imagine walking along what looks like a healthy, meandering stream in the mountains of western Montana. You push through thick growth with high grass brushing your fingertips. All is well until you part the close-set willows and step forward into what looks like the aftermath of a violent explosion."

The land was devoid of life. Skeletons of a few long-dead trees rose out of blue-tinted, powdery dirt. Copper salts had precipitated up from the ground, causing flocking as on a Christmas tree, he said. A bone, they were told, left out to winter on the salts would turn the color of turquoise by spring. Slickens were pockets of concentrated mine tailings washed down by great floods from Butte and Anaconda and contained a poisonous mix of arsenic, lead, copper, cadmium, and zinc. They could be found on both banks of the Clark Fork River down to Deer Lodge.

Slickens weren't the only challenge along the river, however.

"Everywhere I looked the land appeared to be worn out and used up," he wrote. "Noxious weeds grew in the borrow ditches. A glance toward the pastures found more bare dirt than growth. This overwhelming sense of depletion, coupled with the day's tour of the superfund site's toxic smorgasbord, made me wonder, 'Who in his right mind would choose to live here?'"

He did, as it turns out. A year and a half later, Bryce was hired to manage the 3,000 acres and 140 mother cows of the Dry Cottonwood Creek Ranch—home to a substantial slickens site along the river. The coalition had a number of goals in mind. In addition to building bridges to the ranching community, it wanted to participate in a slickens project as a way of field-testing restoration strategies and demonstrating how the impending Superfund cleanup of the river could benefit farmers and ranchers in the area. With miles of heavily polluted stream bank and a prime location in the upper part of the Deer Lodge Valley, the Dry Cottonwood Creek Ranch was a perfect fit. Bryce spent nearly four years on the ranch, working with other ranch hands to accomplish the following:

- They facilitated the Superfund cleanup process both on and off the ranch, working with contractors hired by the state of Montana to integrate a rigorous soil-sampling program with the ranch's agricultural operations.
- They completed an ambitious restoration project involving planting, seeding, riparian fencing, and stock water development on Dry Cottonwood Creek.
- They successfully experimented with the production and local sale of healthy, grass-fed beef. They raised and finished their own calves and marketed them to consumers within the Clark Fork watershed. Though the program only made use of about 10 percent of the calf crop, it was profitable.
- They partnered with schools to bring students to the ranch for hands-on learning about stream restoration and ecology. Students monitored reaches of Dry Cottonwood Creek and then returned later in the semester for a day of tree planting.

- They used electric fencing and a short-duration/high-intensity grazing regimen to improve livestock distribution, wildlife habitat, and the ecological condition of the land.
- They created realistic and profitable models for sustainable ranching. Between 2009 and 2012, they added about $75,000 of new annual income to the ranch's bottom line.
- They secured funding from Keystone Conservation, a nonprofit group that works to mitigate conflicts between wild predator species and cattle, to defray the cost of hiring a herder to manage livestock on their national forest grazing allotment.

By trying to show ranching neighbors that good stewardship of the land, waterways, and livestock could help, rather than hurt, the economic viability of the ranch, and by creating physical and intellectual common ground between ranchers and conservationists, Bryce had officially become a new agrarian.

"The experience of running a ranch like Dry Cottonwood has left me with two lasting sentiments," Bryce wrote. "The first is a terrible sense of urgency, deriving from the fact that the landscape sustaining us is arid, finite, and fragile. For a long time we have diminished it annually, and we must now find ways to stop. The second thing is an insistent, durable hope that projects like Dry Cottonwood could help us turn an essential corner in the way we approach the stewardship of the West."

We must find ways to make our living from the land in perpetuity, Bryce continued, without exhausting the wild systems that feed our individual and cultural souls. Solving this big riddle, however, is a tricky proposition. It requires old knowledge, new ideas, innovative technology, energy, courage, and lots of *querencia*.

"Projects like the Dry Cottonwood Creek Ranch embrace a collaborative, constructive ethic," he concluded. "This ethic directs us to fix rather than to fight. It counsels us to learn instead of litigate. If pursued at length it can create and unite a community of ecologically altruistic ranchers and pragmatic conservationists. It can turn a landscape—even a hard-used one like the poisoned upper reach of the Clark Fork River—into a place worth calling home."[11]

Silicon + Carbon = Technology for Us All

AN AGRIVOLTAIC RESEARCH FARM, MONTPELLIER, FRANCE

It only took one glance.

While attending a conference in Germany, I snagged a freebie publication titled *Adaptation Inspiration Book*, created by a group called CIRCLE-2, which stands for Climate Impact Research & Response Coordination for a Larger Europe (I'm not sure what the "2" stands for). It was a little book, full of photos and case studies from around the continent that highlighted local climate change adaptation strategies. There were stories about flood abatement projects, rain gardens, coastal defense works, green roofs, heat warning systems, and agroforestry research. Good stuff, if not entirely mesmerizing.

It took only one look at the photo on page 141, however, to cause a double take.

It was a picture of a typical vegetable farm, with nice, neat rows of lettuce, under a blue sky. It's what stood above the field that caused the double take. Between the lettuce and the sky was a bank of *solar panels*. They rested on a sturdy wooden frame 10 feet or so high and cast narrow shadows across the vegetables—shading them! From the heat. From the rising effects of climate change. Making renewable energy. I turned the page. Another photo peered down at the lettuce through the solar panels from up high. The panels sparkled in the sunlight. The farm was growing food while simultaneously making electricity!

I felt a jolt.

That's because for a while now I've despaired over a seemingly intractable question: what is the best way to utilize sunlight—to grow food or to produce fuel?

For millennia, the answer was easy: we used solar energy to grow plants that we could eat. Then in the 1970s the answer became more complicated as fields of photovoltaic panels (PVPs) began popping up around the planet, sometimes on former farmland. This was part of a new push for renewable energy sources, and as the technology has improved over the years so has the scale of solar power projects on land that could otherwise produce food.

In the 1990s, the food-versus-fuel debate took a controversial turn when farmers began growing food crops for fuels such as corn-based ethanol, with encouragement in the form of government subsidies. Today the production of biofuels, including massive palm oil plantations, has become big business, often at the expense of hungry people. As a result, the land requirement of the biofuels industry, not to mention its deleterious impact on ecosystems and biodiversity, has become huge—and it keeps growing.

Making the situation even more complicated and controversial is a simple fact: according to scientists, the amount of land needed to replace fossil fuels with biofuels exceeds all the farmland available on the planet. In other words, increased competition between food and fuel for agriculturally productive land means that the stage is set for food shortages and rising conflict as the projected human population on Earth swells to nine billion by 2050.

This food-versus-fuel debate has drawn considerable attention recently, with a number of potential solutions being proposed as a way out of what is quickly becoming a serious conundrum. Here are a few, briefly:

- **It's a definition issue:** According to a group of researchers, the trouble is that no one can agree on what defines "surplus" land, including idle, marginal, reclaimed, and degraded land. Devising a common language, they say, means we'll be able to "creatively utilize surplus land" for energy and the environment.
- **It's a plant issue:** Researchers say we're using the wrong feedstocks for bioenergy production. Native grasses, flowers, and herbs offer the best chance for creating

sustainable biofuels instead. Making that dream a reality, however, would require new technology to harvest, process, and convert this plant material, says one report.

- **It's an ethics issue:** Is it moral to produce fuel from food that could otherwise feed hungry people, and to drive up food prices as a result? Is it right for rich nations to exploit poor ones for their fuel needs? Before we can resolve this conflict, say philosophical types, we need to sort out our ethics first.
- **It's a free market issue:** Allocating which parcel of land should be used for food and which for fuel should only be determined by free market mechanisms, say many in the private sector. And the role of government subsidies and regulations should be minimized, they add.
- **It's a geopolitical issue:** The use of land for food or fuel cannot be separated from wider global struggles for economic security, political dominance, and social justice, say activists and government leaders.
- **It's a technological issue:** The conflict can be solved, say engineers, by improvements in solar technology on the one side and plant productivity on the other. This includes ongoing research to "improve" photosynthesis, a chemical process considered by some biotech companies to be too inefficient (I kid you not).

Notice that all of these options have one thing in common: they still see it as a choice between food *or* fuel, silicon *or* carbon. There is no common ground, no coexistence, no win-win solution.

Except there is—I spied it on page 141! That's why I did a double take. Solar panels above a farm field! When I returned home, I excitedly contacted the scientists behind the farm in the photos. Here's what I discovered:

The food-versus-fuel conundrum led French agricultural scientist Christian Dupraz to ponder a question: could food and fuel production be successfully combined on one plot of land? Specifically, why not build solar panels above a farm field so that electricity and food could be produced at the same time? In addition to resolving the conflict between land uses, he hypothesized, solar panels could provide an additional source of income to farmers while sheltering crops from the rising temperatures and destructive hail- and rainstorms associated with climate change.

"As we need both fuels and food," he wrote in a scientific paper published in 2010 in the journal *Renewable Energy*, "any optimization of land use should consider the two types of products simultaneously."

He said it wasn't a new idea. It was first proposed in a 1982 paper titled "On the Coexistence of Solar-Energy Conversion and Plant Cultivation" by two German scientists. But their idea had never been implemented—until Dupraz and his colleagues at the French National Institute for Agricultural Research (INRA) decided to give it a try.

In the paper Dupraz also coined a new word to describe this system: *agrivoltaic*.

To test their hypothesis, Dupraz and his fellow researchers built the first-ever agrivoltaic farm, near Montpellier, in southern France. In a 2,000-square-meter test field they planted crops in four adjacent plots—two in full sun (as controls), one under a standard-density array of PVPs (as if the solar panels had been mounted on the ground), and one under a half-density array of PVPs. The panels were constructed at a height of 4 meters (12 feet) to allow workers and farm machinery access to the crops.

The main issue, they knew going in, was the effect of *shade* created by the PVPs on plant productivity. The researchers assumed productivity would decline, though there was scant data in the scientific literature to consult. That's why they built two different shade combinations, full- versus half-density—so that they could compare the effects of each to the other and to the control plots in full sun.

"Basically, solar panels and crops will compete for radiation," Dupraz wrote in the paper, "and possibly for other resources such as water, as solar panels may reduce the available water quantity for crops due to increased runoff or shelter effects."[1]

By the same token, shade can improve the productivity of crops in a warming world. Water availability limits many crop production potentials, he wrote, and shade will reduce transpiration needs and possibly increase water efficiency.

As the experiment progressed, it became clear to the researchers that a compromise needed to be struck between maximizing the amount of electricity produced by the solar panels and maintaining the productive capacity of the farm. It was the Goldilocks Principle at work: too much shade hurt the crops, too little hurt electricity generation. Everything had to be *just right*. Could this balance be achieved? Variables the researchers identified included

- the proper angle or tilt of the PVPs
- the proper spacing between solar panels
- making adjustments for localized conditions (such as latitude)
- choosing between fixed panels or panels on trackers (cost is a factor)
- the proper height of the PVP array
- engineering issues involved with the construction of the structure that holds the PVPs in place (must be durable)

By the end of three growing seasons they had their answer: yes, balance was possible. But not quite for the reason they expected.

Not surprisingly, the crops under the full-density PVP shading lost nearly 50 percent of their productivity compared to similar crops in the full-sun plots. However, the crops under the half-density shading not only were as productive as the control plots; in a few cases *they were even more productive!*

The reason for this surprising outcome, according to Hélène Marrou, who studied lettuce in the plots, was the compensating ability of plants to adapt to lower light conditions. She reported that lettuce plants adjusted to decreased levels of radiation by (1) an increase in the total plant leaf area and (2) an increase in total leaf area arrangement in order to harvest light more efficiently.

She also had good news to report on the water front. "We showed in this experiment that shading irrigated vegetable crops with PVPs allowed a saving of 14–29% of evapotranspired water, depending on the level of shade created and the crop grown," she wrote in a 2013 paper (one of three).[2] In the context of global warming and water shortage, she said, reducing water demand by shading plants could represent a big advantage in the near future.

Dupraz noted that while commercially available solar panels operate at 15 percent efficiency, the intrinsic efficiency of photosynthesis is "quite low" at roughly 3 percent (which is why companies are trying to "improve" it). This makes PVP systems more attractive to landowners than farms from a solar radiation perspective, especially at big scales. Combined, however, silicon and carbon systems can be very efficient, he wrote.

Being good scientists, Dupraz and colleagues were careful to say that more research was needed, including questions about rain redistribution under the panels, wind effects on the crops, soil temperature changes, the effect of dust from farming on PVP efficiency, and the validity of the results for various latitudes—and a special focus on plants that have a demonstrated ability to compensate for reduced light conditions.

However, their early results were very hopeful.

"As a conclusion," Marrou wrote, "this study suggests that little adaptation in cropping practices should be required to switch from an open cropping to an agrivoltaic cropping system and attention should be mostly focused on mitigating light reduction and on plant selection."[3]

To this end, Dupraz wrote to me recently to say the next step in their research is to evaluate the advantages of using *mobile* solar panels mounted on trackers. This would allow them to adjust the radiation levels for crops to meet their physiological needs. It will also allow the panels to be tilted to a vertical position during rainfall events, giving the water a chance to fall uniformly on the crops.

I'll wager this is the ticket.

Silicon and carbon, working together. The food-versus-fuel debate is no longer an either/or situation. Thanks to the work of Dupraz and his colleagues, we know that agrivoltaic systems can combine food production with energy production on one parcel of land, while at the same time increasing the resilience of agriculture to climate change.

There was more to my double take, however, than just the image of solar panels above a farm field.

What the researchers *didn't* investigate were solar panels placed above organic, cover-cropped, no-tilled, glomalin-rich farm fields. I wondered: what could an array of PVPs do for Dorn Cox, for example, or the heat stress that Annie Novak struggles with on her rooftop farm? A lot, I bet. What would be the effect of solar panels on perennial crops, or an edible garden forest? What about solar arrays above rangelands? That might sound nutty, but think about the huge solar farms being installed right now around the world in sun-drenched locations. What if they were lifted high enough for, say, Colin Seis to get his combine underneath to harvest the oats cropped into his perennial pasture?

I'm certain someone will raise engineering and agronomic objections to these ideas, saying they're impractical and uneconomic. But then, Seis heard the same objections to pasture cropping. So did John Wick and Sarah Mack to their ideas. They took these challenges as opportunities. After all, don't we love a good challenge? Don't we revel in double takes? Until I saw the photo of Dupraz's agrivoltaic farm, I never imagined such a possibility even existed. Ditto with pasture cropping and organic no-till farming. And repairing damaged creeks with sticks and rocks. Rooftop farms and backyard forests. Restoring river deltas. Building soil carbon. They all seemed impossible until a problem-solver came along and said, "I have an idea."

We're an idea species. And a hands-on species. It's not a coincidence that nearly every one of these innovative practices relies on human technology to go along with nature-made ecological processes. We're creative and physical. When both are engaged, good things can happen, like solar panels above farm fields.

Sometimes, all we need is inspiration.

7

CONVERGENCE

On a sunny late summer day I flew to northern Utah for the last planned stop on my journey through Carbon Country. I went to visit two old colleagues, learn about a new technology, see big country, and ponder measurements, cycles, and markets. And close a circle in my life.

I headed to the famous Deseret Ranch, a beautiful and biologically diverse 215,000-acre property owned by the Church of Jesus Christ of Latter-Day Saints in sagebrush steppe country east of Logan, Utah. Founded in 1891, its abuse over the decades by too many livestock and too little care, and its subsequent rebirth in the 1980s as a pioneer in the emerging conservation ranching movement after its purchase by the church, was an inspiring story for many of us, myself in particular. At the time, I was looking for ranches that demonstrated the core principles of the New Ranch—that the natural processes that sustain wildlife habitat, biological diversity, and functioning watersheds are the same processes that make land productive for livestock. This description fit the Deseret perfectly, thanks to the visionary leadership of Gregg Simonds, its former general manager, and Rick Danvir, the ranch's longtime biologist, both of whom I would be meeting again on this trip.

Gregg had been hired by the previous owner and given the steep challenge of turning the ranch around financially without spending any additional money, a goal that he accomplished. When the Mormon church purchased the property its business leaders were so impressed by the ranch's financial numbers that they decided to keep Gregg in his job, demonstrating the persuasive power of monitoring, as Gregg puts it. The new owners gave the green light to Gregg and Rick to continue to improve the livestock operation's profitability while expanding and maintaining the property's diverse wildlife population. This meshed nicely with Gregg and Rick's belief that livestock and wildlife were codependent on each other economically and ecologically. It was a view that many in the conservation community, especially wildlife specialists, rejected outright. Gregg and Rick knew from separate experiences, however, that time-controlled livestock grazing, brush thinning, and prescribed fire were tools that could restore damaged ecosystems, recalling Aldo Leopold's observation that wildlife "can be restored by the creative use of the same tools which have heretofore destroyed it—axe, plow, cow, fire and gun."

Their goal was to use these tools to create a mosaic of vegetation types and age groups across the ranch. Under previous management, the plant community had been drastically simplified by poorly managed cattle grazing, with unhappy consequences for wildlife. Their plan worked. Creative use of cattle under Gregg's leadership restored the land's ecological complexity while increasing the ranch's stocking rate. Rick directed other habitat improvement work, including prescribed fire and brush-thinning projects. Wildlife populations responded as a result, and the land began to heal.

Numbers backed them up.

In a paper delivered at the Seventieth North American Wildlife and Natural Resources Conference in 2005, Rick reported that wildlife abundance and species diversity had risen over the years and remained high even as the cattle stocking rate continued to increase. The ranch supported over 275 bird species, he wrote, and had been designated an "important bird area" by the Utah chapter of the Audubon Society in 2002. Pronghorn antelope and elk populations had increased under the ranch's management strategy, and densities of sage grouse, an at-risk species, were twelve times greater than on adjacent lands. On the financial side, all range restoration work and management expenses associated with the wildlife program were being covered by increased livestock numbers and hunting income.

"We suggest time-controlled grazing is functionally and aesthetically preferable to either season-long grazing or livestock removal," he wrote in the paper. "Further, managing for a productive system and diverse landscape can be economically self-sufficient and ecologically sound, simultaneously enhancing at-risk wildlife populations and ranching."[1]

It was the New Ranch at work. That's why I had stopped by for a visit in 2004—to see its possibilities at a large scale, and to be inspired.

These were the same reasons I returned nearly nine years later, though this time I had carbon on my mind.

When I came up with the idea of a carbon ranch during the summer of 2009, inspired by Sarah Scherr and Sajal Sthapit's publication *Mitigating Climate Change through Food and Land Use*, I spelled out in an essay six strategies to increase or maintain soil health and thus the carbon content of grass- or shrub-dominated ecosystems of the American West. When I did so, I had the Deseret Ranch in mind. My subsequent travels confirmed my hunches, and I think it's worth revisiting them here as a summary before pushing on.

1. **Planned grazing systems.** The carbon content of soil can be increased by three principal methods: the establishment of green plants on previously bare ground; deepening the roots of existing healthy plants; and the general improvement of nutrient, mineral, and water cycles in a given area. Planned grazing is key to all three. By controlling the timing, intensity, and frequency of animal impact on the land, the carbon rancher can improve plant density, diversity, and vigor. Specific actions include the soil cap–breaking action of herbivore hooves, which

promotes seed-to-soil contact and water infiltration; the "herd" effect of concentrated animals, which can provide a positive form of perturbation to a landscape by turning plant litter back into the soil (an intensive version of this effect is sometimes called a "poop-and-stomp"); the stimulative effect of grazing on plants, followed by a long interval of rest (often a year), which causes roots to expand while removing old, oxidized forage; targeted grazing of noxious or invasive plants, which promotes native species diversity and vigor; and the targeted application of animal waste, which provides important nutrients to plants and soil microbes.

Additionally, planned grazing systems—including management-intensive, time-controlled, short-duration, and mob grazing systems—have the advantage of focusing the practitioner's attention on the day-to-day and week-to-week condition of the land. This enables the manager to achieve specific ecological goals effectively, such as the goal of increased quantity, density, and vigor of green plants (and thus increased carbon storage).

2. **Active restoration of riparian, and wetland areas.** Many arroyos, creeks, rivers, and wetlands exist in a degraded condition—the result of historical overuse by humans, livestock, and industry. The consequence has been widespread soil erosion, loss of riparian vegetation, disruption of hydrological cycles, decline of water storage capacity in stream banks, loss of wetlands, and many other examples of land "sickness." The restoration of these areas to health, especially efforts that contribute to soil retention and formation, such as the reestablishment of humus-rich wetlands, will result in additional storage of atmospheric CO_2 in soils. The "toolbox" for the restoration of these areas is now well developed and practical, and it could be implemented at scale if desired. There are many cobenefits of restoring riparian areas and wetlands to health as well, including improved habitat for wildlife,

increased forage for herbivores, improved water quality and quantity for downstream users, and a reduction in erosion and sediment transport.

3. **Removal of woody vegetation.** Many meadows, valleys, and rangelands have witnessed a dramatic invasion of woody species, such as piñon and juniper trees, over the past century, mostly as a consequence of the suppression of natural fire and overgrazing by livestock (which removes the grass needed to carry a fire). The elimination of overabundant trees by agencies and landowners, via prescribed burns or other means, has been the focus of much restoration activity in the Southwest recently. The general goal of this work is to encourage grass species to grow in place of trees, thus improving the carbon-storing capacity of the soil. Not only can soils store more CO_2 than trees, they also have the advantage of relative permanence. Trees can burn up, be cut down, die of disease or old age—all of which can ultimately release stored CO_2 back into the atmosphere. Additionally, the removal of trees has an important cobenefit: they are a potential source of local biomass energy production, which can help reduce a ranch's carbon footprint.
4. **The conservation of open space.** The loss of forest, range, or agricultural land to subdivision or other types of development can dramatically reduce or eliminate the land's ability to pull CO_2 out of the atmosphere via green plants. Fortunately, there are multiple strategies that conserve open space today, including public parks, private purchase, conservation easements, tax incentives, zoning, and economic diversification that helps keep a farm or ranch in operation. Perhaps most importantly, the protection of the planet's forests and peatlands from destruction is crucial to an overall climate change mitigation effort. Not only are forests and peatlands important sinks for CO_2, but their destruction releases large amounts of stored carbon back into the atmosphere.

Note that "protection" may still result in the loss of stored carbon if the land stewardship practices don't maintain or improve the health of plants and soil. A farm or ranch, for instance, may be protected from development by a conservation easement, but its poor ecological condition (or its poor management) may cause CO_2 to leak back into the atmosphere. This is one reason why those farms and ranches that have already improved the health of their land, and thus improved the carbon storage capacity of their soils, need to be supported economically, socially, and politically, so that their human caretakers benefit from their good work and continue to practice good stewardship.

5. **The implementation of no-till farming practices.** Plowing exposes stored soil carbon to the elements, including the erosive power of wind and rain, which can quickly cause it to dissipate back into the atmosphere as CO_2. No-till farming practices, especially organic ones (no pesticides or herbicides), not only protect soil carbon and reduce erosion, they often improve soil structure by promoting the creation of humus. Additionally, farming practices that leave plants in the ground year-round both protect stored soil carbon and promote increased storage via photosynthesis. An important cobenefit of organic no-till practices is the production of healthy food.
6. **Building long-term resilience.** Nature, like society, doesn't stand still for long. Things change constantly—sometimes slowly, sometimes in a rush. Some changes are significant, such as a major forest fire or a prolonged drought, and can result in ecological threshold-crossing events, often with deleterious consequences. *Resilience* refers to the capacity of land, or people, to "bend" with these changes without "breaking." Managing a forest through thinning and prescribed burns so that it can avoid a destructive, catastrophic fire is an example of building resilience into a system.

> Managing land for long-term carbon sequestration in vegetation and soils requires building resilience as well, including the economic resilience of the landowners, managers, and community members. For example, cooperation among disparate individuals or groups, such as biologists, conservationists, ranchers, and policy makers, with the goal of improving land health, can help build ecological and economic resilience within a watershed. This can have two important effects: direct storage of CO_2 in the soil, as humus is created, and the strengthening of relationships required for the maintenance of healthy soil over time.

To this list I would now add *two new items* plus one *caveat* and one *conundrum*.

The *caveat* is straightforward: In arid and semiarid environments—for example, much of the American West—the carbon cycle works slowly, especially in a drought. This is important, because amidst the general hope about soil carbon and the possibility of the world's rangelands and temperate grasslands sequestering large amounts of atmospheric carbon dioxide, often lost is any mention of the effects of drought on the carbon cycle. I didn't consider them either when I first began to look into this topic. Not only do dry times significantly reduce a plant's vigor and its ability to grow and reproduce, and thus store carbon, they can also increase the amount of CO_2 respiration that takes place at the other end of the carbon cycle—decomposition—especially if a plant or a community of plants *dies*. Active management is key to keeping the carbon cycle functioning as well as possible in dry times, but if there isn't enough forage for hungry animals, then the options available to a rancher may be quite limited. Furthermore, scientists tell us that under climate change conditions future droughts in the region may be longer and deeper, especially in already dry country like the Southwest.

In other words, we must be careful about how much greenhouse gas mitigation we can expect from arid environments going forward.

On the other hand, this caveat means we should redouble our efforts in creating sweet spots—wetlands, peatlands, riparian areas, wet meadows, springs, and other areas of high biodiversity as well as grasslands that get higher amounts of precipitation annually—and not just for carbon's sake. All of these areas contribute multiple, critical ecosystem services to humans and wildlife and should be restored and maintained in good health for this reason. In this way, carbon sequestration can be a cobenefit of overall good land stewardship.

One of the *two new items* to add to my list is aesthetics, which I didn't fully appreciate when I first set out on my journey. I knew from our work healing creeks with Bill Zeedyk that form and function are inseparable, but not until I traveled around Carbon Country, meeting many creative people, did I realize just how interwoven beauty is with healthy land, good work, and positive energy—just as Leopold suggested it would be.

Beauty is in the eye of the beholder, of course, but I'll speculate that there are core principles at work in all of us—the way a sunset resonates, for example, or the sight of a rainbow after a rain shower. What I do know for certain is that you need form *and* function. They reinforce each other like bricks need mortar to make a wall. Beauty can't fix creeks or heal wounds by itself, but at the same time pure functionality won't motivate many people. Success requires both. When it comes to relationships, whether between people, land and people, or elements of land itself, science needs art and art needs science.

Which brings me to the *second added item*: more data.

We need more numbers to support our experience, intuition, and desire for beauty. Specifically, we need more quantitative monitoring of the land—a topic that often makes nonwonky types go glassy-eyed, including me sometimes. That's certainly how I felt when I set out to explore Carbon Country, but I see now that without numbers we can't make much progress. This was the main reason I returned to the Deseret—to see the wonky. From the start of their careers here, Gregg and Rick were determined to gather data as part of the ranch's normal operations. They did it for a very practical management reason: without solid data, you're just taking guesses. Gregg calls such guesswork "faith-based management" and compares it to picking stocks in the stock market by closing your eyes and saying "eeny-meeny-miney-moe." Sometimes our prayers are answered; often they are not. Collecting numbers makes sense for decisions about wildlife, but Gregg considered it absolutely critical to the livestock operation as well. His favorite saying is, "If you can't measure it, you can't manage it," which is why he has dedicated more than a dozen years of his life since leaving his job on the Deseret to developing a scientifically valid but land manager–friendly monitoring protocol for quantifying land health that will be respected by the marketplace. Data is the key, he believes, to creating market value and thus economic incentives to make a carbon economy work. The protocol was almost ready to debut and he wanted to give me a sneak preview.

Which is where Homer Simpson comes in.

I met Gregg and Rick at the Deseret headquarters, and after a round of introductions and some catch-up conversations we headed to the ranch's spiffy hunting lodge, located high in the foothills. The ranch is truly lovely—big, wide-open country rising steadily from rolling sagebrush steppe to subalpine mountains. Deer, elk, black bears, and tons of other wildlife abound on the property, which is why it is such a popular (and profitable) hunting destination for church members and colleagues. I could tell as we drove that the land was still in good condition despite a series of dry years. Rick said wildlife populations were holding steady, which was good to hear. These were all reasons why I had the Deseret Ranch in mind when I drew the top-left corner of my carbon map, adding mountains, wildlife, forests, prescribed fire, thinning projects—and a monitoring crew.

The turnaround job that Gregg accomplished with the cattle operation on the Deseret has become something of legend in the world of progressive ranching, and after he left the ranch he parlayed his experience into a successful consulting career, mostly working with owners of large landscapes. Back in 2004, I caught up with him on a 400,000-acre ranch in northern Nevada where he was beginning to develop a range monitoring program using Landsat satellite imagery, available for free from the federal government, to check on the ecological health of the whole ranch at once. Traditionally, monitoring is done in short surveys, called transects, in a limited number of spots around the ranch—often a very limited number. And not repeated very often. Gregg thought this was a terrible way to gauge the health of a landscape. It was like having a doctor make a diagnosis of your overall health after examining only a toe, an ear, or a part of your arm—once every ten years.

He had a new idea: to correlate the pixels from the satellite images to actual conditions on the ground.

We settled into a corner of the hunting lodge while Gregg fired up his PowerPoint presentation. Gregg put up on the screen a blank image of the 400,000-acre ranch in Nevada and then began to fill it in here and there with little colored squares representing traditional monitoring sites. After a few clicks of the mouse, he asked a leading question: could we tell what's going on ecologically across the ranch from these few squares? We knew the answer: of course not! Gregg clicked more colored squares onto the map. Could we tell now? Nope. More colored squares appeared. Then more. Nope. More appeared and I had a sense suddenly that I was seeing a puzzle being filled in. What was it? More squares followed and then suddenly they tipped into an image—of Homer Simpson! One final click filled in the entire ranch with Homer's bug-eyed mug. We laughed. It was a great way to make a point—and reinforced my feeling about the role of aesthetics in communication, if you can call Homer Simpson beautiful.

Gregg's point was this: if you want a complete picture of an entire ranch, use a monitoring protocol based on satellite data that is subsequently verified and correlated by a crew on the ground. Anything less means going back to faith-based management decisions. Way cool. And very wonky. I won't go into the details of Gregg's methodology, except to say his protocol focused on "greenness" (i.e., plant carbon) and water and could look back in time thanks to decades of Landsat images. This combination of ground-truthing and satellite imagery made it possible to monitor the ebb and flow of the color green across any size landscape with a great deal of accuracy. It was data that would hold up, Gregg said, in a marketplace designed to pay landowners for their management of carbon or water. Use of remote sensing technology like this was quite revolutionary, I realized, and its potential to move management from faith-based claims to empirical verification could be significant. As Gregg hoped, we were inspired by its possibilities.

Which brings me to the *conundrum*: how do we create a marketplace that will pay landowners and others to double the carbon content of their soil?

Think of all the good things that would happen if the carbon content of the world's soils were doubled from 1 percent to 2 percent, or from 2 percent to 4 percent. Think of the abundance that would happen as a result. Consider the amount of food that could be produced on the same stretch of land, or how much water could be stored in the soil. Think about no-till and organic cover-cropping and the amount of life that would be present in the soil if we let mycorrhizal fungi do their glomalin-making thing. Think about all the nutrients that would be available once more to plants and animals and us as a consequence of doubled carbon. Think about the aboveground wildlife that would benefit from a vibrant, diverse, and abundant belowground ecosystem. Think about all the ecosystem services that would be provided to all living things if we doubled the carbon content of our soils. Then think about how much CO_2 we could sequester in the ground. Not first, but last, meaning sequestration as a cobenefit of stimulating life, not the other way round as I had originally considered it.

Think of all the positive energy that would happen.

How would we make this happen economically? We know how to do it ecologically, as I have detailed throughout this book, and thanks to Gregg Simonds and Whendee Silver and others we now have ways to monitor our carbon ranch's progress, which is the last piece of the land management puzzle, I believe. Now how do we get our economy to help?

One answer, of course, is an incentive-based carbon offset marketplace like what Sarah Mack is trying to get going in southern Louisiana, or a compliance-based system guided largely by cap-and-trade mechanisms, such as the one being developed in California. However, these marketplaces are complicated, bureaucratic, and politically vulnerable, as a recent general election in Australia demonstrated when a change in government brought in a new leader who is vowing to dismantle the country's carbon marketplace. And when considering cap-and-trade schemes, don't forget the aggregating sharks—those speculators, investors, and middlemen who insert themselves into the transaction. Additionally, it is proving difficult to get offset money into the hands of farmers and ranchers to compensate them for their carbon work. The money tends to flow toward technological solutions instead, such as energy efficiency, emissions reduction schemes, "green" infrastructure, and the like. Not much, I bet, has made its way into new soil carbon.

Could there be another model? I'll propose one here, modestly, and then jump out of the way.

What if we *paid* farmers, ranchers, and other landowners or managers directly to double the carbon content of their soils?

What if we said to a farmer or rancher, "We'll pay you $100,000 for every 1 percent of soil carbon increase above a baseline measurement"—what would happen? Let's leave the details out for a moment and fantasize about the big picture. If soil carbon had a high value to society and we were willing to pay to have it increased over time, wouldn't a landowner respond? Wouldn't he or she say, "Hell yes, I can do that!" Better yet, if society didn't dictate which tool to use to achieve this goal, via regulation say, and left it up to the landowner to choose, wouldn't the incentive be even greater to give it a go?

What if we said to a landowner: we'll enter into a contract to pay you $200,000 to double the carbon content of your soil in ten years, and how you accomplish this goal is up to you, whether you use cattle, goats, beavers, pasture cropping, solar panels, wetland restoration, edible backyard forests, holistic livestock management, flerds, grass-fed beef, drought-tolerant seeds, milpas, water harvesting, rooftop farms, no-till organic farming, cover crops, spiders, permaculture, satellite imagery, food cooperatives, biodiesel, open-source software, mycorrhizal fungi, nematodes, earthworms, beer, sheep, podcasts, weed dating, ecosystem services, inspirational lectures, or sweaty dancing.

You choose.

It wouldn't matter what they chose because *you can't increase soil carbon with a practice that degrades the land.* The only way to double soil carbon is with practices that are regenerative and make the land healthier. Take cattle, for instance. If you overgraze the range, carbon stocks will fall, not increase. Plants will suffer, roots will wither, and carbon will leak away. To increase soil carbon with livestock, you must manage them in such a way that promotes plant vigor and thus strengthens the carbon cycle, especially in a drought. Ditto with farms and wetlands and backyard forests and riparian areas. Organic no-till farming will increase carbon stocks; plowing will vaporize them. If, after ten years, the carbon content of a farm or ranch's soil has doubled, fulfilling the terms of the contract, then you can feel assured that the landowner got there with sustainable, regenerative practices of his or her own choosing.

The reason is simple: carbon doesn't lie. It is readily measured and quantified, whether by the spoonful or by a satellite. It increases, decreases, or stays the same. It can't be negotiated, fudged, bullied, bribed, denied, or fooled. It's there or it's not. Either you doubled the amount of carbon in your soil, or you did not.

That's the beauty of the idea: offer to pay landowners to double the carbon content of their soil, then stand back as they choose from the regenerative toolbox, knowing that no matter what methods they choose they'll be creating a cascade of cobenefits, including food, fuel, fiber, forage, water, and fun. Better yet, if landowners can get beavers, grass plants, or nematodes to do most of the work, they'll do it for free and never ask for a vacation!

It's all about renewing life. Carbon is life. Grow carbon and you grow life. Do things that encourage life and you'll grow carbon—blue, green, or brown carbon, take your pick.

I know there will be many people itching to raise an objection to this idea. Wait! Wait! What about drought? What about variability in weather patterns? What if carbon stocks fall—does a landowner have to pay the money back? What if a landowner intentionally degrades his or her land just so he or she can bring it back and make money? What about the good stewards—how will they be compensated for the good work they've already accomplished? Whose protocols will be used? That's too much money! That's not enough! It'll never work!

There will be other objections, ones that people won't be itching to shout out loud, but ones that will be just as difficult to overcome. For example, you can't get rich making soil carbon. Not filthy rich. Not wealthy in the Wall Street sense. It won't help you get something for nothing either. It's hard work. Soil carbon doesn't come out of a slot machine or a lottery wheel; it can't be discovered in big, thick seams in your backyard to be mined and sold as a commodity, making you unexpectedly rich; it can't be hawked to you by a corporation, controlled by a government, or cornered by a conglomerate. It'll never have an initial public offering, see a stock split, or pay a dividend—except as healthy soil. There's nothing virtual about soil carbon; you can't Google it, tweet it, or stick an annoying virtual ad on it, unless you can figure out a way to get protozoa to carry microscopic signs.

Soil carbon won't make you a millionaire, which makes it a hard sell in our current economy. Carbon's abundance and ubiquity, much like sunlight, also makes it confoundingly democratic. These are good things, I believe. As the century wears on and the effects of climate change begin to bite, as they have already begun to do, then definitions of wealth, success, and happiness will begin to shift, I'm certain. As the global population surges toward nine billion in just a few short decades, soil carbon will begin to look more and more valuable, not as a way to become wealthy, but as a way of ensuring our well-being. Ditto with adaptation to climate change. Strategies to cope with drought, feed more people, create new habitats for wild animals, store more water in the soil, abate heat waves, floods, and other weather extremes, and adjust to "new normals" in general will depend, at their core, on carbon.

The sooner we get started, the better.

There will be one more objection to my idea: it isn't practical. The toolbox isn't diverse enough, people will say, or big enough to work at scale. Or they'll simply disagree about the regenerative nature of the practices, echoing Colin Seis's moron principle again. More fertilizer! More diesel! More business-as-usual! We need more technology and engineering to solve the rising challenges of the twenty-first century, people will insist, not more fungi. My response is simple: *it is practical*. These practices can work. We can do this at scale. We're an ingenious species, the most ingenious ever in the history of the planet (alas). Give us a problem to solve and some tools to do the job and then stand back.

There is another objection, however, that's 100 percent legit: where will the money come from? If we are going to double the carbon content of America's soils, we're going to need a lot of money. That's because we're talking about a lot of soil. I'm not even going to attempt the math. So where is the money going to come from? I'm not an economist, but two thoughts come to mind.

One, we're a rich nation. A really rich nation. We have tons of money. Maybe we could use part of our vast wealth to double soil carbon, restore degraded watersheds and rangelands, increase biodiversity, lower agricultural greenhouse gas emissions, produce renewable energy, mentor a generation of young agrarians, and grow a lot of healthy local food. Maybe we could steer a tiny portion of the Defense Department's budget to agriculture and conservation in the name of *genuine* national security. Silly me!

Second, we could impose a carbon tax and use a portion of the proceeds to pay landowners to double the carbon content of their soils.

I'm not going to go into the specifics surrounding a carbon tax here, except to say that if we ever get serious about climate change—which we will someday—then a carbon tax is probably inevitable. It's the only strategy that makes sense. And in a way it parallels my carbon payment idea: tax carbon at its source and let market forces respond; make carbon payments available and let landowners respond. A big portion of a carbon tax will be needed to offset the rise in the cost of fossil fuels, probably through a reduction in payroll income taxes or payment as an annual dividend (as they do in Alaska), but a portion could be set aside to fund soil carbon projects, with the goal of doubling carbon in soils in ten years, say. I have no idea how much money would be necessary, but I bet it would be a fairly small percentage of the revenues generated by a carbon tax.

And the poetic justice of using a black carbon tax to fund brown or blue carbon projects would be delicious.

Simple. Tax fossil fuels at a fairly high rate and stand back as the economy shifts to cheaper renewable energy sources. Greenhouse gas emissions would fall. Efficiency would rise. No complicated regulations or rules or mandates for this or that would be necessary. People would adjust. Other taxes would decline. A bunch of money would be generated. It could be used for the communal good, such as doubling soil carbon.

Simple. And hard. That's why it is a conundrum.

As I drove away from the Deseret Ranch after my visit, an idea popped into my head: increasing the soil carbon of our land by 2 percent would take only 2 percent of America's population (the percentage of active farmers and ranchers today) for only 2 percent of the nation's gross domestic product (GDP). In other words, for a small amount of money, a tiny amount of people could double the carbon content of the nation's farms and ranches with a *huge* positive effect (I don't know what it would cost actually in terms of GDP, but it wouldn't be very much). Best of all, the toolbox required to accomplish this important goal already exists! The wide selection of tools now available have been field-tested successfully in a variety of landscapes by hard-nosed types, many of whom I met on this journey through Carbon Country. We just need to put them to work. I decided to call this regenerative toolbox "2% Solutions" and vowed that when I returned home I would begin writing short case studies to help spread the news.

There was another reason to feel optimistic: hope in the form of the new agrarians. Young people understand the issues at work today in a way that those of us who grew up with paradigms aligned for a different era cannot very well. The idea of sustainable farming or ranching, for example, is not the radical proposition it was when I became active in the environmental movement in the mid-1990s (though it still is for some of the die-hard old guard). In fact, the integration of conservation and sustainable food production is one of the great leaps forward in recent years—a development that has been largely led by young people. It's a no-brainer to them. They also bring a great deal of business sense and tech savvy to their work—also no-brainers. This also represents a huge shift.

When I began my Sierra Club work, I was taught that conservation and business were mutually exclusive. The purpose of conservation was to shield nature from profit. At the time, this made sense, evidenced by open-pit mining, clear-cut forests, and overgrazed range. However, eventually new models of use were developed, thanks to pioneers like J. I. Rodale, Allan Savory, Bill Zeedyk, and many others. Organic agriculture, permaculture, no-till farming, and collaborative watershed-based conservation rose and challenged conventional thinking about humans and nature. Much has changed since I stalked the halls of Congress in the mid-1990s looking for votes in support of designating new wilderness areas in the West. I find this development very hopeful. The movement has grown and diversified. The young leaders of new agrarianism stand on strong shoulders.

Young people also bring a can-do zeitgeist to modern challenges that flouts the "leave-it-alone" conservation philosophy of my generation. Humans shouldn't do anything, I heard repeatedly during my activist days, because we'll just make it worse. The idea of restoring damaged creeks, for instance, was a big no-no among my peers, unless it meant removing human use (cattle grazing, for instance) altogether. We were a dark Midas—everything we touched supposedly turned to dust instead of gold. I didn't agree with this philosophy then, and I especially disagree with it today. A medical analogy might be useful. In medicine's early years, ignorance ruled, and as a consequence doctors often made a bad situation worse (such as bleeding people with leeches). Sometimes, the prudent course of action was to do nothing. But not anymore. Today, medicine has made huge leaps and we can now heal a wide variety of ills. It's becoming the same way with nature. What we know about the microbial universe in the soil, for example, is nothing short of amazing. What we know now about carbon is incredible as well. The leave-it-alone option isn't really an option anymore, especially under the ominous storm clouds of climate change. Young people especially understand this. Let's push on, they say.

Of course, I shouldn't be trying to speak for young people here. They're perfectly capable of speaking for themselves. I will say that they're aware of the challenges ahead and many seem undaunted, which is a very hopeful sign. I don't know if they have moments of discouragement, though I suspect they do, and I don't know if they get rattled by our dysfunctional political system or the awesome power of industrialism—hopefully they do not. I don't know if they think about carbon much yet, though some do, certainly. If this recounting of my journey through Carbon Country has helped these new agrarians in any way, I'll have considered my task largely fulfilled. They are the carbon farmers and ranchers of the future. They are the ones who will do the lion's share of the planting, growing, restoring, and harvesting of carbon. They are the ones who will double the carbon content of our soils—if we can figure out how to let them try. They deserve every ounce of support we can give them, and we can start by helping them to stand on our shoulders.

Which brings me to the rest of us.

As I said at the beginning of this book, we are all on the map. Every one of us. We're all carbon. We live in a carbon universe. We breathe carbon, eat carbon, use carbon products, profit from the carbon cycle, and suffer from the carbon poisoning taking place in our atmosphere. If we stopped for a moment to consider the condition of our carbon—is it "good" carbon or "bad"?—we would begin to view the consequences of our actions in a different light. We could, for example, find ways to support the 2 percent of Americans who actively manage the soil portion of the carbon cycle. There are a million ways to help them, starting with the power of the purchasing dollar. Seek out the new agrarians and buy their products. Better yet, get involved yourself. Carbon is everywhere—if we take the time to look.

You live in Carbon Country. We all do. Let's return to where we started and make it a place worth calling home once more.

ACKNOWLEDGMENTS

Every book, like every journey, is a collaboration, and this one would not have been possible without the assistance, creative ideas, and hard work of each individual and organization mentioned in the text. I can't thank them enough. I was propelled into my journey by a quartet of innovators—Sarah Scherr, Christine Jones, John Wick, and Jeffrey Creque—and I am indebted to them for their kindness, patience, and willingness to share their knowledge. Their inspiring work and good humor motivated me to keep exploring, and I am especially indebted to Dr. Jones for making arrangements for me to visit carbon farms Down Under. As I traveled deeper into Carbon Country, I found myself being reinspired over and over by the people I met and the stories they shared with me. I was especially impressed by the cadre of young agrarians I met along the way. It is my fondest hope that this project fires up both the spirits and the actions of young people.

I would not have gone far into this new territory without the tremendous support of the staff, board, and friends of the Quivira Coalition. In 2010, they embraced my far-out idea of a carbon ranch for the theme of our annual conference, which wasn't without controversy in some quarters. They enthusiastically supported the launch of the Carbon Ranch Project the following year, which allowed me to investigate the ideas raised in the conference in depth, and in 2012 I was generously granted a sabbatical from my duties as executive director to pursue this book project. In particular, I am indebted to Avery Anderson, who stepped up to run Quivira in my absence. She did such a great job we decided she should continue as the permanent executive director. I couldn't have accomplished this book without Avery's support, talent, and unquenchable good cheer.

I also couldn't have accomplished this book without the financial support for the Carbon Ranch Project. It costs money to take a journey, as everyone knows, and I am very grateful for the help I received. In particular, the Compton Foundation of San Francisco, California, was instrumental in the success of this project. In early 2009, I took a running leap off a cliff with a grant proposal to Compton in which I proposed to explore the idea of carbon-neutral, sustainable livestock production on rangelands (this was before I read Sarah Scherr's report). To my delight, they funded the grant and the Carbon Ranch Project was born. Compton continued its support over the ensuing years, demonstrating the important role philanthropy can play in incubating new ideas that might otherwise languish. They have my heartfelt thanks. So do the other funders: the Lydia B. Stokes Foundation and the New Cycle Foundation, of Santa Fe, New Mexico; and the Lia Fund, the Panta Rhea Foundation, and the Giles W. and Elise G. Mead Foundation, all of northern California. The Quivira Coalition also provided important financial support, especially during my sabbatical.

The collaboration extended to friends and family, and I am very grateful for all the support I received. Although we didn't get to travel together as we did for my first book, Gen, Sterling, and Olivia were essential to the success of this book, especially for their love and smiles. Thinking about the world that Sterling and Olivia will inherit someday was a primary motivation for my journey, and I hope the book will aid in some way in creating a brighter future for them and their cohort. I am indebted to Ben Watson and the fine folks at Chelsea Green Publishing for taking the publishing leap together. Their enthusiasm spurred me into the last stages of the writing process. I also appreciate the help George Greenfield gave me in shaping the book. And endless hugs to the rest of the Quivira gang for their assistance and encouragement: Tamara, Kit, Catherine, Mollie, Michael, Virginie, and Deanna.

I'd also like to include a small shout-out to Jesse, our dog, who gave me daily excuses to take long walks under ever-blue New Mexican skies, enabling me to productively sort out my tangled carbon thoughts.

Portions of the text were previously published in *Acres*, *Farming*, and *Solutions* magazines. I'd like to thank the editors for their support.

In 2012, I had the honor of being a writer-in-residence at the Ucross Foundation, near Sheridan, Wyoming, as well as the first Aldo Leopold Writer-in-Residence at Mi Casita, in Tres Piedras, New Mexico, courtesy of the Aldo Leopold Foundation and the US Forest Service. These opportunities allowed me to concentrate on this book, and I'd like to thank everyone involved for their support.

Lastly, I'd like to acknowledge those researchers, innovators, and practitioners who didn't make it into the book but whose contributions to improving our carbon universe have been significant on many fronts: microbiologists, organic farmers, food activists, policy wonks, renewable energy experts, range scientists, ranchers, restoration specialists, conservationists, educators, eaters, growers, and so many more. We all stand on tall shoulders. As we move deeper into this uncertain century, we'll need them to see as far into the future as possible. I'm certain this story is just beginning, and I'm already looking forward to the next leg of the journey and the stories I'll discover and want to share.

Thanks again.

NOTES

Prologue

1. "Keeling Curve Lessons: Page 4 of 7"; Scripps Institution of Oceanography's CO_2 Program, University of California–San Diego, 2007, scrippsco2.ucsd.edu/program_history/keeling_curve_lessons_4.html.
2. "G-8 Failure Reflects US Failure on Climate Change," by James Hansen; Huffington Post, 2009, www.huffingtonpost.com/dr-james-hansen/g-8-failure-reflects-us-f_b_228597.html.
3. *Mitigating Climate Change through Food and Land Use*, by Sara J. Scherr and Sajal Sthapit; WorldWatch Institute (report no. 179), Washington, D.C., 2009, www.worldwatch.org/node/6126.

Chapter One: Essence

1. Marin Carbon Project: www.marincarbonproject.org/.
2. "Scientists Help Ranchers Wrangle Carbon Emissions," by Christopher Joyce; National Public Radio, Washington, D.C., December 10, 2009, www.npr.org/templates/story/story.php?storyId=121200619.
3. "Adapting Farming to Climate Variability," by Christine Jones; Amazing Carbon, April 2009, www.amazingcarbon.com/PDF/JONES-AdaptingFarming(April09).pdf.
4. "Grassland Management and Conversion into Grassland: Effects on Soil Carbon," by Richard Conant et al.; in *Ecological Applications* 11(2), 2001, eprints.qut.edu.au/37788/1/cona2282.pdf.
5. "Effects of Organic Matter Amendments on Net Primary Productivity and Greenhouse Gas Emissions in Annual Grasslands," by Rebecca Ryals and Whendee Silver; in *Ecological Applications* 23(1), 2013, www.esajournals.org/doi/abs/10.1890/12-0620.1.

Chapter Two: Abundance

1. "Glomalin: Hiding Place for a Third of the World's Stored Soil Carbon," by Don Comis; Agricultural Research Service, US Department of Agriculture, Washington, D.C., September 2002, www.ars.usda.gov/is/ar/archive/sep02/soil0902.htm.
2. *Conquest of Abundance: A Tale of Abstraction versus the Richness of Being*, by Paul Feyerabend; University of Chicago Press, 1999.
3. "On Building a Regenerative Economy," by Dorn Cox; in *Resilience*, no. 38, 2012, quiviracoalition.org/images/pdfs/5280-Journal38_web.pdf.

Profile: Essential Minerals

1. "Conservation," by Aldo Leopold; in *Round River: From the Journals of Aldo Leopold*, edited by Luna Leopold, Oxford University Press, 1993.

Chapter Three: Coexistence

1. *Communities in Landscapes Project Benchmark Study of Innovators, Gulgong, Central West Catchment NSW*, by Peter Ampt and Sarah Doornbos; University of Sydney, New South Wales, 2010, sydney.edu.au/agriculture/documents/2011/reports/Ampt_CiL_BM6.pdf.
2. For more on Jo Robinson's research into grass-fed beef, see her website: www.eatwild.com.
3. *Greener Pastures: How Grass-Fed Beef and Milk Contribute to Healthy Eating*, by Kate Clancy; Union of Concerned Scientists, Cambridge, Mass., 2006, www.ucsusa.org/assets/documents/food_and_agriculture/greener-pastures.pdf.
4. "NASA Researchers Make First Discovery of Life's Building Block in Comet," by Bill Steigerwald; NASA's Goddard Space Flight Center, news release, August 17, 2009, www.nasa.gov/mission_pages/stardust/news/stardust_amino_acid.html.
5. "DNA Building Blocks Can Be Made in Space," by Bill Steigerwald; NASA's Goddard Space Flight Center, news release, August 9, 2011, reprinted by Space Daily, www.spacedaily.com/reports/DNA_Building_Blocks_Can_Be_Made_in_Space_999.html.
6. Letter from Charles Darwin to Joseph Dalton Hooker, February 1, 1871; in *The Life and Letters of Charles Darwin*, vol. 2, edited by F. Darwin, Basic Books, 1959 (reprint).
7. "From Primordial Soup to the Prebiotic Beach: An Interview with Exobiology Pioneer Dr. Stanley L. Miller," by Sean Henahan; Access Excellence, 1996, www.accessexcellence.org/WN/NM/miller.php.
8. "The Limits of Organic Life in Planetary Systems," by the Committee on the Limits of Organic Life in Planetary Systems, National Research Council; National Academies Press, National Academy of Sciences, Washington, D.C., 2007, www.nap.edu/catalog.php?record_id=11919.

Profile: A Carbon Sweet Spot

1. "Re-establishing Marshes Can Turn a Current Carbon Source into a Carbon Sink in the Sacramento–San Joaquin Delta of California, USA," by Robin Miller and Roger Fuji; in *River Deltas: Types, Structures and Ecology*, edited by David Contreras, Nova Science Publishers, 2011, www.novapublishers.com/catalog/product_info.php?products_id=21836.
2. "Blue Carbon Initiatives Emerging as Promising Carbon Sinks," by Suzanne Bohan; in *Contra Costa Times*, January 23, 2011, www.contracostatimes.com/bay-area-news/ci_17164343?nclick_check=1.

NOTES

Chapter Four: Resilience

1. *Restoration of Degraded Deltaic Wetlands of the Mississippi Delta*, by Sarah Mack et al.; Tierra Resources, 2012, tierraresourcesllc.com/restoration-of-degraded-deltaic-wetlands-of-the-mississippi-delta/.
2. *Ecosystems and Human Well-Being: Our Human Planet: Summary for Decision-Makers*, part of the Millennium Ecosystem Assessment by the United Nations; Island Press, 2005, www.unep.org/maweb/en/index.aspx.
3. "Minimizing Carbon Emissions and Maximizing Carbon Sequestration and Storage by Seagrasses, Tidal Marshes and Mangroves," by Emily Pidgeon et al.; International Working Group on Coastal "Blue" Carbon (report), 2011, www.marineclimatechange.com/marineclimatechange/bluecarbon_recommendations_files/bluecarbon_recommendations_3.28.11.FINAL.HIGH.pdf.
4. *The Ancestor's Tale: A Pilgrimage to the Dawn of Evolution*, by Richard Dawkins; Houghton Mifflin, 2004.

Profile: Leave It to Beavers

1. "Working with Beaver for Better Habitat Naturally! Working with Landowners, Teaching about the Ecological Benefits of Beaver, Building Flow Devices, Live-Trapping and Relocating, and More," by Sherri Tippie; Grand Canyon Trust (report), 2000, grandcanyontrust.org/documents/ut_workingBeaver2010.pdf.
2. *Beaver as a Climate Change Adaptation Tool: Concepts and Priority Sites in New Mexico* by Cathryn Wild; Seventh Generation Institute (report), 2011, www.seventhgeneration.org/files/Beaver_As_a_Climate_Change_Adaptation_Tool_-_Concepts_and_Priority_Sites_in_New_Mexico2.pdf.

Chapter Five: Affluence

1. *Let the Water Do the Work: Induced Meandering, an Evolving Method for Restoring Incised Channels*, by Bill Zeedyk and Van Clothier; Quivira Coalition, 2009.
2. *Revolution on the Range: The Rise of the New Ranch in the American West*, by Courtney White; Island Press, 2008.
3. "Land Pathology" can be found in *The River of the Mother of God and Other Essays by Aldo Leopold*, edited by Susan Flader and J. Baird Caldicott, University of Wisconsin Press, 1991.
4. "The Conservation Ethic," in *The River of the Mother of God*.
5. "The Role of Wildlife in a Liberal Education," in *The River of the Mother of God*.
6. "Deer and Forestry in Germany," a lecture delivered by Aldo Leopold; as quoted in *Aldo Leopold: His Life and Work*, by Curt Meine, University of Wisconsin Press, 1988.

7. "Letter to the Editor" by Aldo Leopold; *Journal of Forestry* 34 (April 1936): 446.
8. *A Sand County Almanac: And Sketches Here and There*, by Aldo Leopold; Oxford University Press, 1949.
9. "Restoring Hózhó: A View from the Back Forty Thousand," by Tammy Herrera; in *Resilience*, no. 35, February 2010, quiviracoalition.org/images/pdfs/2055-Journal35.pdf.

Profile: Seeing the Edible Forest for the Trees

1. *Paradise Lot: The Making of an Edible Garden Oasis in the City*, by Eric Toensmeier and Jonathan Bates; Chelsea Green, 2013.
2. *Edible Forest Gardens* (two volumes), by Dave Jacke with Eric Toensmeier; Chelsea Green, 2005, www.edibleforestgardens.com/.
3. More can be found on Toensmeier's website: www.perennialsolutions.org/.

Chapter Six: Emergence

1. Quotes are from a series of short essays written by Annie Novak for *The Atlantic*: www.theatlantic.com/annie-novak/.
2. Brooklyn Grange's website: www.brooklyngrangefarm.com/.
3. "The Agrarian Standard," by Wendell Berry; in *Citizenship Papers: Essays by Wendell Berry*, Shoemaker and Hoard, 2003.
4. The Greenhorns website: www.thegreenhorns.net/.
5. "Yes We Are Farming: Direct Action for the New Economy," by Severine von Tscharner Fleming; in *Resilience*, no. 38, 2012, quiviracoalition.org/images/pdfs/5280-Journal38_web.pdf.
6. "A Durable Scale," by Eric Freyfogle; in *The New Agrarianism: Land, Culture, and the Community of Life*, edited by Eric Freyfogle, Island Press, 2001.
7. "Sembrando Semillas: Planting Seeds of Traditional Agriculture for Future Generations," by Miguel Santistevan; in *Quivira Coalition Journal*, no. 31, September 2007, quiviracoalition.org/images/pdfs/1578-Journal_31.pdf.
8. "Miguel Santistevan—Mentor, Farmer & Educator—Taos, NM—Part 1 of 3," http://youtube.com/watch?v=xXWCXo-THfg.
9. "Querencia: The Soul of the Paisano," by Estevan Arellano; in *Quivira Coalition Journal*, no. 31, September 2007, quiviracoalition.org/images/pdfs/1578-Journal_31.pdf.
10. *Badluck Way: A Year on the Ragged Edge of the West*, by Bryce Andrews; Simon & Schuster, 2013.
11. "Dry Cottonwood Creek Ranch: Superfund Cleanup and Collaborative Conservation in the Contemporary West," by Bryce Andrews; in *Resilience*, no. 38, 2012, quiviracoalition.org/images/pdfs/5280-Journal38_web.pdf.

NOTES

Profile: Silicon + Carbon = Technology for Us All

1. "Combining Solar Photovoltaic Panels and Food Crops for Optimising Land Use: Towards New Agrivoltaic Schemes," by C. Dupraz et al.; in *Renewable Energy* 36 (10), 2011: 2725-2732, www.sciencedirect.com/science/article/pii/S0960148111001194.
2. "Productivity and Radiation Use Efficiency of Lettuces Grown in the Partial Shade of Photovoltaic Panels," by H. Marrou et al.; in *European Journal of Agronomy* 44 (2013): 54-66, www.sciencedirect.com/science/article/pii/S1161030112001177.
3. "How Does a Shelter of Solar Panels Influence Water Flows in a Soil-Crop System?" by H. Marrou et al.; in *European Journal of Agronomy* 50 (2013): 38-51, www.sciencedirect.com/science/article/pii/S1161030113000683.

Chapter Seven: Convergence

1. "Sagebrush, Sage-Grouse and Ranching: A Holistic Approach," by Rick Danvir, John Haskell, et al.; in *Transactions of the 70th North American Wildlife and Natural Resources Conference*, Wildlife Management Institute, 2005, www.wildlifemanagementinstitute.org/store/product.php?productid=16176.

INDEX

H

I

J

K

L

M

N

O

P

Q

R

S

INDEX

Y

Z

ABOUT THE AUTHOR

A former archaeologist and Sierra Club activist, Courtney White dropped out of the "conflict industry" in 1997 to cofound the Quivira Coalition, a nonprofit dedicated to building bridges between ranchers, conservationists, public land managers, scientists, and others around the idea of land health (www.quiviracoalition.org). Today, his work with Quivira concentrates on building economic and ecological resilience on working landscapes, with a special emphasis on carbon ranching and the new agrarian movement.

His writing has appeared in numerous publications, including *Farming Magazine*, *Acres USA Magazine*, *Rangelands Magazine*, the *Natural Resources Journal*, and *Solutions Magazine*. His essay "The Working Wilderness: A Call for a Land Health Movement" was published by Wendell Berry in 2005 in his collection of essays titled *The Way of Ignorance*.

Courtney is the author of the book *Revolution on the Range: The Rise of a New Ranch in the American West* (Island Press, 2008), and he coedited, with Rick Knight, *Conservation for a New Generation*, also published by Island Press in 2008.

In 2012, he published a collection of black-and-white photographs of the American West in an online book titled *The Indelible West*, which includes a foreword by Wallace Stegner.

He lives in Santa Fe, New Mexico, with his family and a backyard full of chickens.